1 MONTH

FREE
READING

at

www.ForgottenBooks.com

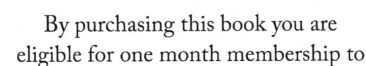

By purchasing this book you are eligible for one month membership to ForgottenBooks.com, giving you unlimited access to our entire collection of over 700,000 titles via our web site and mobile apps.

To claim your free month visit:

www.forgottenbooks.com/free53337

English
Français
Deutsche
Italiano
Español
Português

www.forgottenbooks.com

Mythology Photography **Fiction**
Fishing Christianity **Art** Cooking
Essays Buddhism Freemasonry
Medicine **Biology** Music **Ancient
Egypt** Evolution Carpentry Physics
Dance Geology **Mathematics** Fitness
Shakespeare **Folklore** Yoga Marketing
Confidence Immortality Biographies
Poetry **Psychology** Witchcraft
Electronics Chemistry History **Law**
Accounting **Philosophy** Anthropology
Alchemy Drama Quantum Mechanics
Atheism Sexual Health **Ancient History**
Entrepreneurship Languages Sport
Paleontology Needlework Islam
Metaphysics Investment Archaeology
Parenting Statistics Criminology
Motivational

ISBN 978-1-330-46523-3
PIBN 10053337

HOW TO TEACH
ARITHMETIC

A MANUAL FOR TEACHERS
AND A
TEXT-BOOK FOR NORMAL SCHOOLS

BY

JOSEPH C. BROWN
HEAD OF DEPARTMENT OF MATHEMATICS, HORACE MANN
SCHOOL, COLUMBIA UNIVERSITY

AND

LOTUS D. COFFMAN
DEAN OF COLLEGE OF EDUCATION, UNIVERSITY OF MINNESOTA

CHICAGO   -   -   NEW YORK
ROW, PETERSON AND COMPANY

CONTENTS

PART ONE

PART TWO

PART THREE

PREFACE

This book was written for the purpose of improving the teaching of arithmetic. That arithmetic is poorly taught is indicated by the fact that a larger percentage of children fail in it than in any other subject. The experience of the authors in training prospective teachers and in institute work confirms them in the opinion that the subject is suffering partly because many teachers lack instruction in its theories, methods, and devices. The authors do not assume that method can be substituted for scholarship, but they do contend that teachers as a class want and need definite advice in the teaching of arithmetic. The demands of reading circles, of special methods classes in normal schools and in normal high schools, and of supervisors, but more especially the needs of the classroom teacher, were kept in mind in the preparation of this book.

The number of pages given to the discussion of a topic is no index of the relative importance of the topic in the curriculum; it rather indicates whether a subject is well or poorly taught. However, some chapters are elaborated at greater length than others for the reason that their material is intrinsically or historically more interesting.

This book was not written to exploit any particular textbook or theory or method. It is a simple and accurate exposition of the best methods employed in teaching the subjcet today.

The book consists of three parts: Part I treats the history of arithmetic and the contributions recent scientific studies have made towards standardizing the subject; in

Part II there appears a treatment of certain fundamental principles and ideas that apply to arithmetic in general; and in Part III the methods involved in teaching the various topics or divisions of the ordinary text-book in arithmetic are described in detail.

The authors gratefully acknowledge their indebtedness to Professor David Eugene Smith, of Columbia University; Mr. L. C. Lord, President of the Eastern Illinois Normal School; Mr. E. Fiske Allen, of the State Normal School at Emporia, Kansas; Mr. G. P. Randall, Superintendent of Schools, Danville, Illinois; and Miss Lillian Rogers, of the Horace Mann Elementary School, New York City, for reading certain chapters and for making valuable suggestions.

J. C. B.

L. D. C.

HOW TO TEACH ARITHMETIC

PART ONE

CHAPTER I

THE HISTORY OF ARITHMETIC

Value of a Knowledge of the History of Arithmetic

Chapters and books showing the historical development of a subject of study are frequently given scant notice and indifferent attention, especially by young teachers. Such an attitude, however, never leads to a scholarly knowledge of the subject. With the demands becoming increasingly insistent that teachers know the material they teach several years in advance of their students, information beyond the lids of the text-book in use is regarded as positive evidence of a teacher's professional interest. It is true that a wide margin of scholarship can be secured by studying the more advanced phases of pure mathematics. It is also true that a familiarity with the historical evolution of any subject gives a perspective that enables one to avoid errors, reveals methods and modes of practice that have proved useful or futile, and teaches direct, economical ways of attacking the problems of the subject. In arithmetic we should learn the folly of accepting hasty and unwarranted conclusions and should acquire a method of attacking directly and logically a set of conditions.

In these days of educational agitation, the newer subjects are regarded by many as having the greater educational value. The measure of the educational value of a subject of study is not found in its recency of origin. That subject which has persisted throughout a comparatively long period of time and which is essential to the ordinary life of people over the widest extent of space is the most fundamental. Such studies are not subject to rapid change; their growth is slow. During their evolution they have ministered to the satisfaction of the common needs of the common people. It is for this reason that they are called the *common* branches. Some branches are more common or fundamental than others, for they show a greater persistence in time and in space. No wild appeal of any educational reformer can make the average citizen—no matter whether he lives in San Francisco, New Orleans, New York, London, Paris, Berlin, or St. Petersburg,—believe that knowledge and skill in arithmetic are not necessary to his ordinary life. This feeling is deep-seated and ineradicable.

Not every arithmetical experience has been saved for the present generation to practice upon. Those that were trivial and insignificant, or were of use only to special groups, were soon discarded. Most of those that were saved were the important experiences. Their importance rested upon the degree of their serviceableness to the common people. Those found in the text-books of to-day represent the important experiences, the great typical experiences that have been preserved out of a multitude of attempts at environmental adjustments.

The Slow Growth of Arithmetic

Arithmetic was not born in a day; it is not the child of some fertile mind; it is not the creation of any man, nor of any given group of men, nor of any sect or class in society. It did not spring fully organized into being because groups of educational seers decided on some bright day that the children of the earth needed a new study, and they would organize it in the light of certain *a priori* principles. On the contrary, it had a long period of evolution. It grew by piecemeal. Its beginnings are enshrouded in the mysteries of unwritten history. Its genesis can be traced to-day in some of the practices of primitive man. Of this we are reasonably sure: it was born out of necessity and its more or less disorganized fragments were used for strictly utilitarian ends. It was not until relatively late in the history of civilization that arithmetic became one of the philosopher's chief tools; and it was not until the Middle Ages that it became a handmaiden of theology. Under the influence of the philosophers and the monks arithmetic was divorced from its utilitarian sanctions and became almost wholly a subject for pure speculation.

Why Arithmetic Has Been Taught

Such history as is available seems to warrant the conclusion that most people studied arithmetic and sought its extension, not for any peculiar metaphysical or mind- training value that it might have, but because it was needed in the practical affairs of life. Certainly one of the important reasons for teaching arithmetic was that it was needed in the trade life of the common people. For this reason it was frequently urged by men prominent in Grecian and Roman life that the common people should be instructed in only the simple elements of calculation.

Another claim for arithmetic was that it sharpened the intellect. Curious problems, replete with catch phrases, were ingeniously constructed by the monks as material for disputations. One needs only to glance through educational journals, periodicals, and mathematical magazines to find evidence of the persistence of this mediaeval tendency. It was presumed that arithmetic had peculiar value in training one to think clearly, quickly, and systematically. But this value was to be secured by making the subject difficult, by making it obscure and unnecessarily analytical.

The third reason for giving instruction in arithmetic was the conventional and cultural value of the subject. Solon, Plato, Aristotle, and Pythagoras saw that arithmetic was more than mere calculation. Pythagoras believed that only through the study of arithmetic could one become perfeet and fit for the society of the gods. The Humanists of the Middle Ages, although they added nothing to the subject, upheld the claim that arithmetic was worthy of mastery for its own sake; i. e., irrespective of any use to which it might be put.

The significant fact remains, however, in the face of all cultural, conventional, sentimental, and disciplinary values that may be set upon it, that arithmetic has flourished among trading peoples and has found its real justification there. Wherever a new gateway to trade has been opened up, wherever a new center of commerce has been established, there arithmetic has taken on new life. It is essentially a commercial subject. It has accompanied the march of civilization. When commercial life swept from the Orient into Europe, arithmetic accompanied it; when Italy became the chief trading country of Europe, arithmetic had a commercial sanction there; when trading life became common among the Teutons of the north, arithmetic followed. Practical arithmetic never made progress in the hands of the scholastics or the philosophers.

Arithmetic, like every other subject of study, has evolved slowly. Many of the details of its origin are hidden in unravelled history. We know that during the days of antiquity the Babylonians, the Hindus, and the Egyptians, made rather an elaborate use of number forms and relations. The Babylonians had a knowledge of arithmetical and geometrical progression and perhaps of proportion. It is quite probable that they used the abacus. The sexagesimal system which they used may have been derived from the division of the year into 360 days. The oldest treatise on mathematics that has been deciphered was written by Ahmes, an Egyptian, about 1700 B. C., and was based upon one believed to date as far back as 3400 B. C. This manuscript contains a number of arithmetical problems.

Arithmetic Among the Greeks

The Greeks divided the science of arithmetic into two parts, one of which was called arithmetica and the other logistica. Arithmetica treated the theory of numbers, which were believed to have certain mystical values and were studied and classified as amicable, deficient, perfect, and redundant. Logistica was the art of calculating; mechanical devices such as the fingers, counters, and the abacus were commonly and extensively used in performing its operations. These devices later disappeared with the appearance and introduction of Arabic or Hindu numerals. Dr. Smith thinks that, with the disappearance of these counting devices, there was a corresponding loss of power of real insight into number, and that this loss remained almost wholly unrecognized until Pestalozzi re-introduced object teaching.[1] Naturally, the Sophists gave much attention to logistica, while Plato considered it a vulgar and childish art. Arithmetica was the tool of the philosopher

[1] Smith, "The Teaching of Arithmetic," p. 58.

while logistica was the tool of trade life. Consequently the philosophers looked somewhat with disdain upon the servile tool of tradesmen.

Arithmetic Among the Romans

The Romans contributed practically nothing to arithmetic, unless it be their cumbersome system of notation which has already passed into disuse. Mathematicians speak of Roman life as the period of mathematical sterility. No European country had a better opportunity for standardizing arithmetic and giving certain methods universal validity than the Romans. Although they had great practical ability for organizing armies and colonial dependencies, they seemed to be hopelessly wanting in those imaginative and speculative powers so essential to true scholarship. Moreover, the Romans made almost no attempt to change the language, customs, and traditions of their colonial possessions. The quadrivium consisted of arithmetic, music, geometry, and astronomy. Arithmetic was kept alive through the efforts of such scholars as Boethius, Isidore, St. Boniface, and Alcuin. The ecclesiastics used it to compute Easter and other movable feasts, and for training in disputation. The problems devised for this latter purpose represent a superlative display of mathematical ingenuity.

The Hanseatic League

The Hanseatic League, although it was organized in the thirteenth century to protect trade routes of interest to certain cities, soon took other functions, among which was the general improvement of commerce. To further its purposes it established Rechen Schulen. The Rechenmeisters organized a Guild of Rechenmeisters which was largely instrumental in keeping arithmetic out of the schools until it was introduced by Pestalozzi.

The Renaissance

The Renaissance of the fifteenth and sixteenth centuries received a rich inheritance in arithmetic. During this time the contributions of earlier nations to the various fields of learning were eagerly sought after, and received a hitherto unknown recognition. Life along every line was receiving a new impetus, commerce not excepted. Shipping and ship-building became powerful and inviting trades. They called forth the organization of great stock companies. Business life likewise became collectivistic. Equation of payments came into general use and many elaborate problems in partnership were constructed. Here again there is conclusive evidence of the tendency of arithmetic to conform to the shifting conditions of trade life. ''The sixteenth century gave us printed arithmetic, with important transitions: The transition from the use of the Roman symbols and methods to the use of the Hindu symbols and methods; from arithmetic expressed in Latin to arithmetic expressed in the language of the reader; from arithmetic in manuscript to arithmetic in the printed book; from arithmetic for the learned to arithmetic for the people; from arithmetic theoretic to arithmetic practical; and from the use of counters to the use of figures.'' [1]

Arithmetic Since the Renaissance

Advancement since the Renaissance has been even more rapid. Professor Smith considers progress to have been made mainly along six lines: (1) a marked evolution of the commercial phases of arithmetic, (2) the invention of

[1] Jackson ''The Educational Significance of Sixteenth Century Arithmetic,'' Teachers' College, Columbia University, p. 24.

common symbols of operation probably between 1550 and 1650, (3) the invention of decimal fractions about 1600, (4) the invention of logarithms by Napier in 1614, (5) the modification and improvement of methods of multiplying and dividing and the use of the Austrian method of subtracting and multiplying, (6) the introduction and improvement of algebraic symbols, with the introduction of which certain subjects such as alligation, series, rule of three, and roots either disappeared or began to disappear from the arithmetics, because of the greater ease with which they could be solved by algebra.[1]

Racial Development of Fundamental Phases of Arithmetic

Counting was the initial step in the development of arithmetic by the race. The earliest possible glimmering of the race's intelligence of arithmetic must have come with the counting of like things. This one to one correspondence is the fundamental concept of arithmetic. Computation consisted of nothing more than the simple counting of like things until the race recognized that one object might be considered the equivalent in value of several. Then and not until then did arithmetic really begin. Some light is shed upon the rudimentary character of the number sense by a study of the attainments of primitive peoples to-day. The almost universal tool used by them for counting is the fingers. Whenever sticks, splints, pebbles, shells, notches in sticks, or knots in strings are used as practical methods of numeration, their common origin is explained by the universal finger method. The use of the fingers probably accounts for 10 being the basis of our number system.

Just as the race communicated by oral language before it wrote, so it counted before it invented a system of nota-

[1] Smith, "The Teaching of Elementary Mathematics," pp. 66-68.

tion. One of the most interesting chapters in the history
of arithmetic is that which treats the various systems of
number symbolism. A great number of systems were invented. One can almost trace the intellectual development
of the race in the growth of systems of notation. It is well
known that primitive man proceeded beyond 5 in counting
with the very greatest difficulty. Beyond that he used
such terms as much, many, a heap, or plenty, instead of
definite terms referring to quantities in a series.

The evolution and extension of a series of numbers into
a system was not possible until some number had been
agreed upon and accepted as a base. Only here and there
did the savage mind grasp this fundamental notion. It
was his customary practice to illustrate each successive step
with a particular object. If it could not be illustrated
satisfactorily, he simply referred to it as *many* or *a heap*.
So long as his counting did not exceed the number of fingers on his two hands a system was unnecessary. But when
it became necessary to think ten and one, and to supply
some independent term for the idea, ten became the basis
of his number system. Of course, in the case of the
quinary system the base was established at 5, and in the
vigesimal system at 20. Other numbers have served as
points of departure for the establishment of systems, for
example as in the binary system of Liebnitz, the ternary
and quaternary systems of the Haida Indians of British
Columbia. There is probably no recorded instance of a
number system formed on 6, 7, 8, or 9, as a base.[1]

The Writing of Numbers

The symbolic representation and meaning of the terms
in the different systems varied greatly among different na-

[1] Conant, ''The Number Concept,'' Macmillan, 1896, p. 119.

tions. The Egyptians had symbols for 1, 10, 100, and the higher powers of 10; the Babylonians had symbols for 1, 10, and 100; the Greeks indicated the numbers by giving the initial letters of their names; the Romans resorted to the letters of an old Greek alphabet. All of these plans failed to survive, because they did not contain a symbol to represent zero or naught. This greatest of all mathematical inventions came from the Hindus. It is sometimes erroneously ascribed to the Arabians, but recent investigations show rather conclusively that the Arabs borrowed this idea as well as most of their notary scheme from the Hindus. The invention of zero and its application gave place value to the number series, a fundamental fact that was largely absent from all previous schemes.

Common Fractions

Perhaps the third great step in the evolution of arithmetic following those of counting and notation, was the invention of simple fractions. There is convincing evidence that they were quite generally used at an early date. It seems logical to infer that number ideas, and a number language of integral quantities and relationships, must have preceded in time a system of fractional enumeration. Centuries were required before any adequate or satisfactory system was evolved. The cumbersome character of the early methods is described more at length in the chapter on common fractions. The Babylonians kept the denominator (60) constant and changed the numerator; the Romans kept the denominator (12) constant, and changed the numerator; the Egyptians and the Greeks kept the numerator constant and changed the denominator. Common fractions were in vogue centuries before decimal fractions were invented.

Decimal Fractions

The last fundamental tool for arithmetical work to be invented was decimal fractions. They did not appear until the sixteenth century. They were used as early as 1610 and sporadically thereafter until the eighteenth century, when they were included for the first time as a part of the regular school work, although they were not currently used until the nineteenth century.

Counting, notation, common and decimal fractions, make possible all arithmetical operations. Perhaps the order in which they evolved is indicative of the order in which they should be taught. At any rate, it is indicative of the relation which they bear to one another.

Method in Arithmetic

Instruction in arithmetic was based upon object teaching until the sixteenth century when the Hindu-Arabic numerals were introduced. The introduction of these numerals was followed by a marked improvement in calculation, and a consequent neglect of the philosophy of arithmetic. Text-books were filled with a multitude of rules and examples. Naturally, the work became extremely formal and mechanical. It would be interesting, if not valuable, to trace the persistence of traditions for routine work established then in the changing character of text-books and methods down to the present time. It is certain that arithmetic is still taught in some schools by the committing of rules and the mere solving of problems.

Eventually a reaction set in against the deadening uniformity in whose clutches method was held. One of the earliest attempts to break it up and to reduce the drudgery involved in learning rules, was the publication of rhyming arithmetics. These books became very common during the seventeenth century.

During the long, dark period of teaching without objects, the characteristic mode of instruction was individualistic. Pupils received personal help from the teacher whenever it was necessary. They copied their own problems and solved them in silence. There was little or no class work. There is no way of knowing whether such personal, individual instruction called forth more ability and produced a higher quality of attainment than that produced by the present modes of group instruction, but that it had its inherent strength and advantages is evidenced by the present tendency to reduce the size of classes.

Von Busse

Gottlieb von Busse, 1786, was probably the first to use number pictures (Zahlenbilder) systematically. He arranged them according to this plan:

5

five

It was a short and easy step from the size of the number pictures to a liberal use of objects.

Pestalozzi

Pestalozzi was the first to recognize, appreciate and utilize the full value of objects in arithmetic. He did not approve of the abacus or of geometric forms in teaching arithmetic. The re-introduction of objects meant not only the restoration of the teaching of numbers before figures, but it also operated to free children from slavish bondage to rules. The great concern and reverence of Pestalozzi for childhood expressed itself in a definite attempt to adapt instruction to the mental powers of children. To him the only way to the pedagogical heaven was through object teaching. This resulted in a new and increased emphasis upon oral arithmetic. The subject increased in popularity until it secured the most prominent place in the curriculum.

Tillich

Of the disciples of Pestalozzi perhaps Tillich was the most noted. He continued to simplify the methods and materials of his great master (1) by placing greater emphasis upon the relations of the number forms 1 to 10; (2) by showing that any number of two digits might be considered as so many tens and so many units; and (3) by inventing a Reckoning-chest, which contained 10 one-inch cubes, 10 parallelopipeds two inches high and one inch square, 10 three inches high, and so on to those ten inches high. By means of these blocks he taught notation and the relation of tens and units. Mental arithmetic was made the basis of his work. He exercised more wisdom than Pestalozzi in the selection of suitable materials for perception, and in recognizing the decimal idea.

Frederick Kranckes

In a book by Frederick Kranckes, which appeared in 1819, there is advocated for the first time the teaching of arithmetic by the concentric circle plan. He recommended four circles, the first to contain the number relations from 1 to 10, the second from 1 to 100, the third from 1 to 1000, and the last from 1 to 10000. Like Busse, he made extensive use of number pictures. His book was of a more distinctly pedagogical character than those of his contemporaries, for he planned for the development of the rules and principles of arithmetic through intellectual instruction. In this latter respect he improved upon the teachings of Pestalozzi, whose oral work was, like that of the two preceding centuries, still absurdly formal and abstract.

Grube

Grube (1810-1884) seized upon Kranckes' idea of concentric circles, but reduced the scheme to two circles; the first of which included a year's work on the relations from 1 to 10. This spiral plan became the basic organizing principle of many arithmetics in America.

Although Grube used objects more elaborately than any of his predecessors, he evolved no new principle governing their application. The only new idea he advanced was that all the fundamental operations should be taught in connection with each number before the next number was taken up. In the seventeenth century each topic was completed in turn. Now we have the other extreme, all the number processes taught together before the next number is introduced.

The two chief criticisms against Grube have been his too liberal use of objects and his advocacy of the simultaneous mastery of all the fundamental processes. His point of view as to the second of these is neither natural nor psychological. It is not natural because it does not accord with the racial evolution of these processes nor with the relative difficulties inherent in the various operations. It is not psychological because it is a distinct attempt to reverse the pedagogical maxim, "We must proceed from the simple to the complex."

Object Teaching Over-done

No experienced teacher or supervisor questions the value of object teaching in the primary school. It is not only desirable but necessary that the pupil should be aided in his grasp of number by approaching it from the concrete. However, any plan or method becomes detrimental to the best interests of the pupil if it be over-emphasized. We are frequently reminded that the educational pendulum oscillates between extremes. This has been true in object teaching in arithmetic. Some enthusiasts have carried object teaching to absurd extremes and as a result the pupils have been retarded rather than aided in their mastery of number relations.

The writers are familiar with a city system where the plan of teaching all number ideas by measures and pictures was carried to its logical extreme. The children through the fourth grade were required to solve all problems by concrete demonstration or pictorial illustrations. A comprehensive examination of these same pupils was made when they reached the sixth grade, and by comparing the results with those obtained from sixth grade children in other schools, the conclusion was unmistakable that the children taught by this method had lost almost a full year. They had been arrested upon the plane of concrete thinking in arithmetic. They had not learned to think in symbols, nor had they acquired an automatic mastery of the fundamental operations. We should be open to criticism if we generalized upon a single case, but the results in this instance have been corroborated by superintendents elsewhere. Instead of the method producing better mathematicians, the pupils taught by it knew less arithmetic, were less facile and accurate in computation than other children of equal age, maturity, and training. This fundamental defect

in the claims made for it, we believe, accounts for its
failure.

The Purpose of this Book

The most cautious as well as the most able authors have
recently been attempting to organize the extreme claims of
earlier theorists into a rational scheme.   It is true, the
expressions, "the laboratory method and correlation," are
sometimes used in discussions on the teaching of arithmetic,
but these terms connote methods as applicable to the teaching of other subjects as to arithmetic.   It is the intention
of the authors of this book to gather up these stray threads
of method and to relate them to the best teaching practice
of the day.

CHAPTER II

SCIENTIFIC STUDIES IN ARITHMETIC

There is a tendency in education for experimental study to supplant dogmatic statement. Hitherto progress has been made by the laborious process of trial and error. What seemed to be successful experience has been the criterion for judging practice. But within the last decade a restless discontent with the use of opinion as the final standard for evaluating methods and materials has invaded every field of human activity. The breakdown of old authoritative forms of control has been .accompanied by an increased interest in experiment, investigation, and science. The prevailing tendency to manage commerce, business, and manufacturing scientifically, has expressed itself in education in the attempt to establish units and scales for measuring educational products. An elaborate statistical technique is now employed by the specialist in education in making his investigations and in preparing his scales. It is not our purpose to discuss the nature of this technique nor the reliability of the results secured. Comparatively few of these studies have been made in the field of arithmetic, but these few have illuminated the whole field.

Summary of Rice's Investigation

The first attempt to measure and to evaluate the teaching of arithmetic was made by Mr. J. M. Rice,[1] in which he sought answers to three questions: What results should be accomplished whenever a subject is incorporated in the

[1] Forum 34: 281-297; 437-452.

school program? How much time shall be devoted to the branch? Why do some schools succeed in securing satisfactory results with a reasonable appropriation of time, while others cannot get reasonable results with a satisfactory appropriation of time?

Mr. Rice's test, given in 1902, consisted of a series of eight problems, given to 6000 pupils in the fourth to eighth school years inclusive. The basis used by Mr. Rice in selecting these problems and the manner in which he scored the manuscripts were not explained fully enough to enable other investigators to duplicate the study. However, he arrived at the following conclusions:

1. Variation in ability as shown by the ratings given individual children is common to all grades, but it is greater in the seventh and eighth year classes than in the earlier ones. The range in ability in the seventh year is from 8.9 to 81.1%, and in the eighth year from 11.3 to 91.7%.

2. Superiority, mediocrity or inferiority in any grade in a given city in general means superiority, mediocrity or inferiority in all grades.

3. Differences in home environment do not explain these differences in attainment. Of the eighteen schools examined, three in particular were representative of the ''aristocratic'' districts, and the best of these ranked tenth, while the others ranked eleventh and sixteenth respectively. The school that ranked seventh was located in a slum district. The fifth school in rank was in a district that is but a ''shade better than those of the slums.''

4. Differences in attainment are not explained by differences in size of classes. In general the number of pupils per class was large in the schools that ranked highest, and small in the schools that ranked lowest.

5. Results in arithmetic are correlated with maturity;

that is, the averages improve from grade to grade. That the mere fact of age is not sufficient to account for differences in results is shown by the following table:

	FOURTH GRADE		FIFTH GRADE		SIXTH GRADE	
	Average age	Per cent	Average age	Per cent	Average age	Per cent
Six highest schools......	11.9	62.8	12.6	84.3	13.4	96.3
Six lowest schools.......	11.0	29.0	12.0	49.8	13.4	61.4

	SIXTH GRADE		SEVENTH GRADE		EIGHTH GRADE	
	Average age	Per cent	Average age	Per cent	Average age	Per cent
Five highest schools.....	13.4	49.5	14.1	71.9	14.1	90.4
Five lowest schools......	13.4	11.0	13.11	29.0	14.5	38.0

6. The results indicate that the time of day at which the tests were given was not responsible for the differences in grades received.

7. There is no direct relation between time and results. "The amount of time devoted to arithmetic in the school that obtained the lowest average, 25 per cent, was practically the same as in the one where the highest average, 80 per cent, was obtained. In the former the regular time for arithmetic in all grades was forty-five minutes a day, but some additional time was given. In the latter the time varied in the different classes, but averaged fifty-three minutes daily.

"From these few facts two important deductions may be made: First, that unsatisfactory results cannot be accounted for on the ground of insufficient instruction; and second, that the schools showing the favorable results cannot be accused of having made a fetish of arithmetic. These statements are further justified by the fact that the four schools which, on the whole, stood highest, gave practically the same amount of time to arithmetic as the three schools which stood lowest.''

8. The amount of home work required is no criterion of the results to be expected in school.

''By far the greatest amount of home work in arithmetic was required in the city whose schools obtained the poorest results.'' On the other hand, the five cities standing highest had practically abandoned home work.

9. Methods in teaching are not the controlling element in the accomplishment of results.

In all schools tested by Mr. Rice thoroughly modern methods were used.

10. Variations in results cannot be accounted for by differences in the general qualifications of teachers.

The results in a given city were nearly all good, mediocre or bad; extreme variations did not appear among the different classrooms of the same locality, although the teachers differed greatly in the amount of training they had received.

.11. Mr. Rice reaches the final conclusion that the supervisor is the controlling factor determining differences in achievement in arithmetic. Although one of the primary functions of a supervisor is the preparation of a course of study, it was clearly evident that the excellence of the course of study would not enable one to prophesy as to the achievements of the pupils. The important factors by which the supervisor accomplishes these results are the standards and tests that he uses. In general, the immediate and most potent device used by supervisors in those schools that ranked best was an examination to test the teacher's progress. These examinations emphasized both the reasoning and routine sides of arithmetic. They were suggestive, and stimulated teachers to secure better results.

"Summary of Stone's Investigation"

Dr. C. W. Stone's study on "Arithmetical Abilities and Some of the Factors Determining Them"[1] is one of the most significant investigations of a scientific character. This study was a definite attempt to find the nature of the product of the first six years of arithmetic, and the relation between certain distinctive procedures and the resulting abilities. Dr. Stone personally collected data from twenty-six representative school systems, distributed over New England, the Middle States, and the Central West. Two series of tests were given to the 6A grade in each school system examined, one in fundamentals and one in reasoning.

In these tests given by Dr. Stone a time limit of twelve minutes was set for the solution of the problems in fundamentals, and fifteen minutes for those in reasoning. The time allotted in each case was too little for even the brightest pupil to solve all of the problems. This plan does not correspond with current practice in the holding of examinations. The customary practice is to give a few problems with the expectation that every one will solve them all. This gives some measure of the proficiency of the class, but it does not give a measure of the individual abilities of the pupils. The rating of individuals as to their facility in handling fundamentals or ability to solve problems involving reasoning is best shown when no one has time to solve all of the examples or problems in the test. One such test would show individual differences better than no test at all, but the average of several tests would give a

---

[1] A thorough comprehension of this study can be secured only by an examination of the entire tables. Erroneous impressions will thus be prevented. Published by Teachers' College, Columbia University, New York City.

much more accurate rating of the individuals of the class. Examinations of the kind Dr. Stone used will come into more common use in proportion as teachers become more skillful in evaluating methods and results.

Tests of this kind lose much of their value unless there is a uniform system for scoring the results. The standards used to-day vary with localities, with the training and experience of the teacher, with her inclinations, temperamental attitudes, and personal knowledge of children. One teacher grades on method, so much for trying, another on the correctness of the result; one takes into consideration neatness, and another, punctuation. Obviously there are many factors operating; to speak of standards is really anomalous; we cannot get them until we agree to use uniformly certain common units. No matter how much we may question Dr. Stone's method of checking results, if we wish to compare any school with his we should use the same method. In addition a score of one was given for each column added correctly, and in multiplication a score of one was given for each correct partial product, and for each column added correctly. This same plan was pursued in scoring the examples in subtraction and division. Two methods were used in scoring the problems in reasoning: after the problems had been arranged in the order of relative difficulty, a score was given for each problem reasoned correctly on the basis of its relative difficulty or only the first six problems, which all children had time to solve, were scored.

The problems were arranged as to their relative difficulty by giving one hundred sixth grade pupils time enough to solve all of the problems they could in the order printed.

The problems were next given in reverse order to one hundred pupils who had as much time as they wished to solve them. The average showed the order of difficulty.

What the Scores Measure

"As used in this study the achievements, products, and abilities, except where otherwise qualified, must necessarily refer to the results of the particular tests employed in this investigation. That some systems may achieve other and possibly quite as worth while results from their arithmetic work is not denied; but what is denied is that any system can safely fail to attain good results in the work covered by these particular tests. Whatever else the arithmetic work may produce, it seems safe to say that by the end of the sixth school year it should result in a good ability in the four fundamental operations and the simple every-day kind of reasoning called for in these problems. It does not then seem unreasonable, in view of the precautions previously enumerated, to claim that the scores made in the respective systems afford a reliable measure of the products of their respective procedures in arithmetic." Stone, p. 19.

Relative Difficulty of Column Addition

It is usually assumed that addition increases in difficulty in proportion as the columns increase in length. Dr. Stone found that "one step in a problem in fundamentals is about equal to another, be the step long or short; e. g., 96 per cent of the children did the first column of problem one correctly (six numbers to the column), and exactly the same per cent did the longer and presumably harder column of problem four correctly (eight numbers to the column). Similarly, about as many pupils failed in the very short additions of the partial products in the multiplication problems as failed in the long columns of the multiplication problems." (p. 16, 17) The difficulty that children are supposed to have with columns of extra length is more fanciful than real. Of course, it must be remembered that these tests were given to upper sixth grade children only. It seems quite probable that children of less maturity and training would find columns of different lengths varying in difficulty. This is one of the simple problems awaiting solution.

Arithmetic and Formal Discipline

Practically every scientific study in the field of education has shed more or less light on the mooted question of formal discipline. Although men familiar with the changed point of view have grown increasingly careful in their discussions of the educational value of the various subjects of study, still such terms as arithmetical ability, historical ability, and literary ability are current. To find out whether such a general ability exists in arithmetic was one of the problems of Dr. Stone. Logically, when one uses such terms as arithmetical ability or historical ability, he presupposes that but one ability exists for each of these subjects, that there is no such thing as a plurality of abilities. This conception harks back to the days of faculty psychology, which considered the human mind as made up of separate faculties, each of which was located in a separate compartment of the brain and could be trained by a single subject of study. For example, the faculty of reasoning could be trained best by the study of mathematics, memory by the study of language, and observation by nature study and science. This made education a relatively simple matter. How many subjects should the curriculum consist of was answered for the educationist of that day: just as many as there were faculties of the mind— no more and no less. The advocates of formal discipline went one step farther: they held that skill in one field meant that this skill was correspondingly serviceable **in** every field, that the training in reasoning one got from the study of mathematics made him equally good in reasoning everywhere. The training in memory secured from the study of languages meant an equally tenacious memory in every field. In other words, the training and skill which one secured in any special field could be generalized and applied to every field. A good reasoner in mathematics would, therefore, be equally good in reasoning in metaphysics and theology; a good memory for poetry would likewise mean an equally strong memory for faces, names, dates in history, or scores in games. This effect of special training was another concept which made education a simple matter. Only a few subjects were needed,—just as many as there were faculties of the mind, and each of these had the special and unique function of training some particular ability which in some mysterious way spread itself and made unnecessary the development of this ability in other fields. A few things prepared one for success in any walk of life. It was a beautiful scheme. The value of every subject was interpreted in terms of its mind-training value.

Experience has furnished us with numerous instances of the failure of the plan to work. Modern psychology with its functional point of view has corroborated these common sense experiences by demonstrating conclusively that our mental life consists of abilities and not of faculties, and that these abilities become specialized. Training in one field does not necessarily transfer at all to other fields. This is particularly true if the fields are dissimilar. That doctrine, therefore, which calls for the establishment of generalized habits is a psychological absurdity. Habits are specific responses to specific stimuli. Instead of there being a faculty of attention, a faculty of memory, a faculty of reasoning, there are abilities for attending, for memorizing, and for reasoning, each of which must be trained through responses to a special type of stimulus. The result is that we have those habits of attending, of memorizing, and of reasoning, which correspond to the special fields in which they have been trained. This means that we must teach each fact or theory worth teaching as if the salvation of the intellectual world depended upon it, for it may be that the limited training one gets from any one of them will fail to modify us in some other desirable way. This means that a newer and heavier obligation rests upon the teachers of to-day.

However, we must not presume that our educational forefathers were altogether mistaken. Scientific studies have shown that training in one field is serviceable in another in proportion as the two fields are identical in subjcet matter or in method, or as the ideals of work gathered in one are useable in the other. As many different types of training are needed as there are fields to which students should be adjusted.

Dr. Stone's thesis had certain important contributions to make to this problem: These are shown in the

1. Ratings of Cities

The various school systems measured show a great lack of uniformity in both fundamentals and reasoning. The order of the cities when distributed for fundamentals is not the same as when distributed for reasoning. Not only does the order differ when these two types of arithmetical processes are compared, but it differs when one fundamental is compared with another. For example, City XX, choosing at random, stood lowest in addition, next to lowest in subtraction, third from lowest in multiplication, and fourth from lowest in division. Now, "if the net result of arithmetic were a product, each system would have the same relative position in each phase of the subject.'' (p. 24)

The ratings of cities with reference to the mistakes made, naturally show the same thing; i. e., a plurality of abilities. The twenty-six systems measured range in mistakes made from 14.4 per cent, to 45.1 per cent. The relative rankings of the cities as to mistakes show almost no uniformity at all. Cities are not only unlike in their mathematical achievement, but in their lack of achievement as well.

2. Achievements of Pupils as Individuals

When individuals are compared as to their ability along any of these lines, the wide variability of their achievement becomes obvious. This is true whether we compare individuals in different systems or individuals in the same system. This difference in attainment must be due more to differences in original nature than to differences in the course of study or in the character of the instruction pupils receive.

School systems and individuals differ less widely in their achievements in fundamentals than in reasoning. This may be due to the fact that the fundamentals are better taught, that we know more about the psychology of habit formation than we do about the psychology of reasoning. It seems true that if a class is drilled upon any one of the fundamental operations the variability decreases, whereas if it is trained on problems in reasoning the variability increases. In other words, we are more alike, or may be made more alike, as to our habits in any ability in arithmetic than we are, or can be, as to our abilities in reasoning.

Dr. Stone made a comparison of the attainments of boys and girls with the result that there was no evidence to show that girls are either more or less stupid than boys in arithmetic.

3. Relationship of Abilities

It has already been noted that systems and individuals do not occupy the same stations in different traits in arithmetic. If they did, then the relationship would be perfect; a person, then, would do equally well in each phase of arithmetic,—a certain degree of ability in fundamentals would call for a corresponding degree of ability in reasoning. The extent to which kinship exists between abilities, the extent to which power is transferred from one ability to another, is measured by "the coefficient of correlation." The coefficient of correlation is "a single figure so calculated from the individual records as to give the degree of relationship between the two traits which will best account for the separate cases in the group. In other words it expresses the degree of relationship from which the actual cases might have arisen with the least improbability. It has possible values from + 100 per cent through 0 to — 100 per cent. (Quoted from Thorndike's "Educational Psychology," p. 25) "A coefficient of correlation between two abilities of + 100 per cent would mean that the best system or pupil in the group in one ability would be the best in the other; that the worst system or pupil in the one would be the worst in the other; that if the individuals were arranged in order of excellence in the first ability and then in the order of excellence in the second, the two rankings would be identical, that the station of any pupil in one would be identical with his station in the other. A coefficient of — 100 per cent would, per contra, mean that the best system or pupil in the one ability would be the worst in the other, that any degree of superiority would go with an equal degree of inferiority in the other, and vice versa.'' (Stone, p. 37.) A coefficient of + 62 per cent would mean that any given station in the one trait would imply 62 hundredths of that station in the other. A coefficient of — 62 per cent would mean that any degree of superiority would involve 62 hundredths as much inferiority, and vice versa.[1]

From what has already been said it is reasonably clear that equality of achievement is not secured. The coefficients of correlation between reasoning and the fundamentals bear this out. The twenty-six systems when related show the following correlations:

Addition with subtraction.................	.92
Addition with multiplication...............	.95
Addition with division....................	.90
Subtraction with multiplication............	.95
Subtraction with division.................	.93
Multiplication with division...............	.92

This high correlation between the fundamentals is what we expect. It is of interest to note that the lowest relationship is between addition and division, and that the relationship between addition and multiplication is just the equal of that between subtraction and multiplication.

Comparing the achievements of 500 individuals selected at random, Dr. Stone found the relationships between reasoning and the fundamentals to be as follows:

Reasoning with all the fundamentals........	.32
Reasoning with addition...................	.28
Reasoning with subtraction................	.32
Reasoning with multiplication.............	.34
Reasoning with division...................	.36

[1] The statistical technique involved in the calculation of these coefficients may be found in Thorndike's ''Mental and Social Measurements.'' The Science Press, New York City.

The highest correlation, it will be noted, is between reasoning and division. This probably means that ability in division is a better measure and criterion of ability to reason than is ability in any other fundamental. If one could choose only one fundamental to discover whether children could reason or not, he should choose division.

The achievements of these 500 individuals in the various fundamentals were related as follows:

Addition with subtraction..................... .50
Addition with multiplication................ .65
Addition with division..................... .56
Subtraction with multiplication............. .89
Subtraction with division................... .95
Multiplication with division................ .95

When compared with the table showing the relationships between systems as to fundamentals it is clear that individuals are more unlike as to their abilities than systems are as to their results. This table also lends weight to the hypothesis that the increasing kinship as shown by the correlations is due to the increase in the amount of reasoning involved. "It seems safe to say tentatively of the fundamentals that the possession of ability in addition is the least guarantee of the possession of ability in others; that the possession of ability in multiplication is the best guarantee of the possession in others; and that this probably means that multiplication is like addition on its mechanical side and like division on its thinking side. Hence, if it is desired to measure abilities in fundamentals by a single test, one in multiplication would be best; and a test in division would probably be the best single measure of arithmetical ability." Stone, p. 42.

This array of facts furnishes sufficiently conclusive evidence for the statement that there is no such thing as a product in arithmetic,—what we get is a complex of products. There is no such thing as general arithmetical ability,—there are many arithmetical abilities.

Other Evidence

Dr. Stone's conclusion has been corroborated by several other studies. Professor Daniel Starch,[1] of the University of Wisconsin, conducted a somewhat similar investigation. "Eight people practiced fourteen days on oral multiplication. Before and after practice they were given six tests in arithmetical operations, and two in auditory memory span. For comparison seven other observers were given the preliminary and final tests without the practice series. The practiced observers showed from twenty to forty per cent more improvement in the arithmetical tests than the unpracticed observers." The subjects were required to record their introspections. These seemed to show that improvement was due, not to an increase in the memory span, but to an ability to keep the numbers better in mind. "The improvement in the end tests was due, therefore, to the identical elements acquired in the training series and directly utilized in other arithmetical operations. The two main factors were (a) the increased ability to apprehend and hold numbers in mind, and (b) the acquisition of the ability to utilize arithmetical operations."

Supervision and Results in Arithmetic

A practical question for every community is, "Does supervision pay?" As the field of education is differentiated, supervision by various specialists becomes more

[1] "Transfer of Training in Arithmetical Operations." *Jour. of Ed. Psych.* 2: 306-310.

and more imperative. Apparently some one whose vision reaches to every corner of the field is needed, that the claims and demands of the department heads of a school may be organized for effective work. School organization is one of the essential prerogatives of the school administrator. Does the work of this educational engineer modify achievement in any particular subject? If so, how much and in what ways? As yet no satisfactory answer, based upon trustworthy experimental studies, has been secured to these questions. The twenty-six school systems which we have been describing were checked with reference to whether the superintendent alone supervised the work in arithmetic, the principal alone, or the principal and superintendent combined. Those schools that ranked best were examined by both of these officers; the schools that ranked next best were tested, supervised, and inspected by the superintendent alone; and those that ranked poorest by the principal alone.

Time in Relation to Results in Arithmetic

Our school programs have been arranged in part to correspond to the work curves of children. In some books on school hygiene it is claimed that the work curve of children rises rapidly during the forenoon, reaching its highest mark of the day between ten and eleven o'clock; that it descends until after the noon hour, when it again rises, reaching its high mark between two and three o'clock, after which it gradually descends until bed time. Such difficult subjects as arithmetic and grammar were placed on the program at those hours of the day when the energy of the children was supposed to be at its highest pitch. Subjects like music and drawing were taught when the children's energy was at ebb. This theory, for it was no more than that, was widely taught and applied. Dr.

Stone has given us some new light upon the validity of the theory. He gave his tests at different hours of the school day. The results showed that the children did as well one time as another. In this he corroborated the conclusions reached by Mr. Rice. If teachers fail to get as good results one hour of the day as another in arithmetic, the failure is probably no criticism upon the children. The failure is more probably due to the inability of the teacher to supply adequate motives. The feeling of incompetency may be broken through under the pressure of strong motives and new intellectual levels reached. As far as possible a teacher should be permitted to arrange her program so that she has each subject coming at the time of day when she thinks she can teach it best. Then if she fails the criticism of her work may with justice be far more severe than if she arranges a program to correspond to hypothetical work curves.

What is the relation of the total amount of time expended in arithmetic to efficiency in the subject? If one school gives one hundred minutes a week, another two hundred, and another three hundred, may we expect the results in the second school to be twice as good as in the first, and the results in the third school to be three times as good as those in the first? The fact is the thirteen schools that received less than the median time cost did slightly better than those that received more than the median time cost. In other words, there was no direct ratio between time expenditures and abilities. "A large amount of time spent on arithmetic is no guarantee of a high degree of efficiency." (p. 62.) The chances are exactly even that if one were to choose at random a school having less than the median, or middle, time cost, it would stand among the leaders in arithmetical achievement, and the chances are even that if one were to choose a city

with more than the middle time cost it would rank among
the poorest in arithmetical achievement.

The best results were secured in those cities that had
their time most evenly distributed, grade by grade. Mr.
Rice found that home study did not help matters. Dr.
Stone said that those systems that required home work
got better results than those that did not. This is a
problem demanding further investigation.

Relation of Course of Study to Arithmetic

A good course of study does not guarantee ability in
arithmetic. Some cities with excellent courses of study
were among the poorest in results, while some cities with
poor courses of study were among the best in results. It
seems that ability in fundamentals is more closely related
to the course of study than ability in reasoning.

If efficiency in arithmetic cannot be measured in terms
of the time of day at which the subject is taught, the
amount of time expended upon it, or the general excellence of the course of study, what is its determining cause?
All these things are factors, but the most important factor, aside from supervision, is the teacher who can breathe
personality into the dead materials, and who can transmute
them into the consciousness of the children. How this can
best be done is still largely an unsolved problem. It seems
that there must be many ways in which the lifeless materials of arithmetic can be infused with life. No adequate
survey of them is in print.

Standards in Arithmetic

The most comprehensive as well as the most significant
attempt to standardize achievements in arithmetic is that
of Mr. S. A. Courtis, of Detroit, Michigan. Mr. Courtis

evolved from Stone's study the idea of preparing a series of tests by which one could measure with considerable exactness the development of arithmetical ability through the school and the actual attainment of any individual at any time. This is the kind of tool every progressive teacher and efficient supervisor has been yearning for; and now that it has been established after a tremendous amount of patient labor, its value depends upon the application that is made of it.

The tests prepared by Mr. Courtis are serviceable for both teachers and supervisors. Pupils also can use them for comparing their ability at one time with their ability at another time. No high degree of technical skill or elaborate knowledge of arithmetic is necessary for their application.

Mr. Courtis limited his tests to simple exercises in the four fundamental operations and to one- and two-step problems in reasoning. The tests, eight in number, are as follows:

Test No. 1. Addition
Test No. 2. Subtraction
Test No. 3. Multiplication   Combinations 0-9.
Test No. 4. Division
Test No. 5. Copying figures (rate of motor activity)
Test No. 6. Speed reasoning (simple one-step prob-
             lems)
Test No. 7. Fundamentals (abstract examples in the
             four operations)
Test No. 8. Reasoning (two-step problems)

To measure the various arithmetical abilities represented by these tests grade by grade, the tests must be given in all grades under exactly similar conditions. The tests are printed on separate sheets of paper; the pupils turn these face-up at a given signal; and then work at full speed under a time limit. There are more problems in each test than any pupil can solve in the time given. By this means the teacher gets not only the range and central tendencies of each ability measured, but she has each individual station in the entire distribution. Room can thus be compared with room, grade with grade, age with age, sex with sex, one individual with another. If the results secured for a given ability are secured by different methods and if the pupils in two rooms are of about equal ability, then the tests give a measure of the value of the methods. If the methods are the same and the children of two rooms of equal ability, then the results measure the efficiency of the teaching. By the use of these tests any careful and able investigator could answer once for all the mooted question of whether children in rural schools are better in arithmetic than town and city children. An unpublished study by Dr. E. H. Taylor, of the Eastern State Normal School at Charleston, Illinois, shows that, contrary to expectation, children in the city schools of Charleston and Mattoon, Illinois, and in the training school of the Normal School, without exception averaged higher in the fundamental arithmetical operations than did the pupils in the corresponding grades in the rural schools of the county. It is hardly fair to assume that Dr. Taylor's results are typical for the United States, and yet the authors feel that it is safer to accept them than to accept judgments based upon crude opinion. Whenever experimental evidence is weighed against mere opinion, the burden is upon those who hold the opinion to disprove the facts arrived at through experiment.

In confirmation of Stone's thesis, Mr. Courtis found that school systems differ less widely than the individuals within a system. This, Mr. Courtis believes, "can only mean that the differences in the abilities of individual children are greater factors in determining relative rank in school work than all the differences in abilities of teachers, courses of study, or methods of work combined.'' This, of course, is a matter which can be settled by an extended investigation. The true answer will always be a matter of conjecture until children are taught by the same methods and by teachers of equal ability. Perhaps we could get no nearer to the correct answer than to give the tests to children who have been for a considerable number of years in an orphan asylum; there the educative environment is perhaps more uniform than in any public school system.

Mr. Courtis found that the grades overlapped greatly in each of the abilities measured. This fact is clearly shown in the table given below.

GENERAL AVERAGE FROM TOTAL DISTRIBUTIONS

School Grade No.	Average of scores for each test	Test No. 1	Test No. 2	Test No. 3	Test No. 4	Test No. 5	Test No. 6 (*)	Test No. 6 (†)	Test No. 7 (*)	Test No. 7 (†)	Test No. 8 (*)	Test No. 8 (†)
1	55	6	6	—	—	29	—	—	—	—	—	—
2	75	21	12	10	12	51	—	—	—	—	—	—
3	525	26	19	16	11	63	2.8	2.1	5.4	1.7	2.7	0.6
4	1222	33	25	23	21	70	3.7	2.5	6.6	3.6	2.6	0.8
5	1177	40	32	30	28	80	4.4	3.4	9.0	5.3	2.8	1.2
6	1282	46	37	34	35	88	5.1	4.4	10.3	6.9	3.4	1.7
7	1432	51	40	38	38	98	5.9	5.2	11.5	7.6	3.7	2.2
8	1370	57	45	43	44	102	6.8	6.1	13.1	8.9	4.1	2.7
9	412	59	47	45	47	108	6.9	6.4	13.7	9.5	4.1	3.1
10	216	57	45	43	46	112	7.2	6.7	14.0	9.5	4.1	3.1
11	151	59	47	44	48	114	7.9	7.4	14.4	9.4	4.5	3.3
12	169	61	48	44	49	112	7.7	7.2	14.9	10.8	4.6	3.6
13	462	71	56	50	56	116	8.6	8.2	16.8	12.6	5.3	4.0
14	131	74	51	58	59	124	9.7	9.1	17.2	11.8	5.4	4.1

8679

* Average number problems attempted.   † Average number right.

"This table means, when put into words, that in the third grade, for example, there are 525 individual scores in each test, or about 525. If in some test there were more, as for example, addition, there were enough less in division to make the average number of tests 525. In addition tests it was found that the standard or average ability was that exhibited by a child who could record 26 in a minute. The standard in subtraction was 19 per minute, in multiplication 16, division 11, in copying figures 63, in speed reasoning the pupils attempted, on the average, 2.8 problems and got 2.1 right, and so on.

Mr. Courtis holds that a higher score than the average is desirable for a standard. He, therefore, created a standard by taking the lowest score of the best 30 per cent of the children in each of the eight grades measured. This gives the following table:

STANDARD SCORES

	Test No. 1	Test No. 2	Test Nos. 3 and 4	Test No. 5	Test No. 6		Test No. 7		Test No. 8	
					Ats.	Rt.	Ats.	Rt.	Ats.	Rt.
Grade 3	26	19	16	58	2.7	2.1	5.0	2.7	2.0	1.1
Grade 4	34	25	23	72	3.7	3.0	7.0	3.3	2.6	1.7
Grade 5	42	31	30	86	4.8	4.0	9.0	4.9	3.1	2.2
Grade 6	50	38	37	99	5.8	5.0	11.0	6.6	3.7	2.8
Grade 7	58	44	44	110	6.8	6.0	13.0	8.3	4.2	3.4
Grade 8	63	49	49	117	7.8	7.0	14.4	10.0	4.8	4.0
Grade 9	65	50	50	120	8.6	7.8	15.0	11.0	5.0	4.3

Translating this table into words: "At the end of a year's careful work an eighth grade child should be able to copy figures on paper at the rate of 117 figures per minute; to write answers to the multiplication combinations at the rate of 49 answers per minute; to read simple one-step problems of approximately 30 words in length and decide upon the operation to be used in their solution at the rate of 8 problems a minute with an accuracy of 90 per cent; to work abstract examples of approximately 10 figures (twice as many for addition) at the rate of 14.4 examples in 10 minutes with an accuracy of 70 per cent; to solve two-step problems of approximately 10 figures at the rate of 5 in 6 minutes with an accuracy of 75 per cent. At the present time 70 per cent of the eighth grade children cannot meet these demands. But it must be borne in mind that three per cent of the fifth grade children can, and that experience has shown that individual care and a little well arranged drill produces marked changes in the ability of most children.

Professor Franklin Bobbitt translates this table in this way: "In simple addition operations, the third grade teacher should bring her pupils up to an average of 26 correct combinations per minute. The fourth grade teacher has the task, during the year that the same pupils are under her care, of increasing their addition speed from an average of 26 combinations per minute to an average of 34 combinations per minute. If she does not bring them up to the standard 34, she has failed to perform her duty in proportion to the deficit; and there is no responsibility upon her for carrying them beyond the standard of 34. Her task is simply to increase their addition rate from 26 to 34. The fifth grade teacher is to take pupils with an average rate of 34 and bring up their speed to an average of 42, a perfectly definite task. The sixth grade teacher is to take pupils with an average of 42 and to carry them before the end of the year to an average of 50 combinations per minute. The seventh grade teacher increases their ability from 50 combinations to 58. The eighth grade teacher takes them with 58 combinations per minute and brings them up to 63, and the ninth grade teacher is to add the small increment of 2 combinations per minute during the ninth grade. In like manner, in the case of each of the other operations, each teacher has his own special increment to add to the work of his predecessor before turning his partially finished product over to the next teacher in the school. This table of standard scores of Mr. Courtis shows us the ultimate standard that is to be attained at the end of the school course, and it also shows the progressive standards to be attained at each stage of the process from the beginning to the end.'' [1]

Reasoning in Arithmetic

The reasoning ability of children in the fourth, fifth, and sixth grades in the public schools of Passaic, New Jersey, was tested by Dr. F. G. Bonser. [2]

Perhaps the most significant single contribution made by this study was the confirmation of President G. Stanley Hall's theory of the periodicity of the learning process. It was found that progress is not regular, but that there is a well defined rhythm. The nodes or crests of the wave for boys appear at about 9 years and 6 months, at about 12 years, and at about 14 years and 6 months. ''For the girls, there is evident, though not so clearly, a rhythm with its crests about coincident with the valleys of the rhythm for the boys, excepting at the period 11 years 6 months to 12 years 6 months, where the crests become nearly parallel. In so far as these tests and these children are typical, then we can predict that of any group of children of these grades, ranked on the basis of ability, a larger proportion of those pupils who are from 12 to 13 will be found in the highest group than of those who are from 11 to 12.''

---

[1] Quoted from ''The Twelfth Yearbook of the National Society for the Study of Education. Part I, pp. 21-22.''

[2] The Reasoning Ability of Children. Teachers' College, Columbia University, Contributions to Education, No. 37.

His results also showed that the greatest gains irrespective of sex were made in the 4A and 5B grades; the smallest gains for boys in 5B and 5A grades and for girls in the 6B and 6A grades. His results also confirmed the opinion that those of superior ability were the youngest age group. Because of this it is easily possible that unless the school is flexible in its promotions the brightest pupils may be the most seriously retarded. A comparison of the sexes showed that boys are a little better than girls in reasoning ability in arithmetic, and slightly more variable.

The studies that relate more specifically to practice and practice effects will be discussed in the chapter on Drill Work.

PART TWO

CHAPTER III

ACCURACY

Prevalence of Inaccuracy

A common criticism of the schools of to-day is that the pupils have been permitted to become lax and careless in thought and in expression. The modern pupil is expected to study many things which were not taught in the schools of the last generation, but there is truth in the assertion that no small part of his knowledge is superficial and inaccurate, "a collection of vague ideas rather than clear cut notions about definite things." It is asserted that while the pupils of to-day can think and write on more subjects than could the students of former years, their expressions are less clear and coherent. Teachers should not turn a deaf ear to the complaints about inaccuracy of thought and statement. The complaints come from many sources and seem to be corroborated by overwhelming evidence. The employer maintains that it is difficult to hire a boy who is accurate in statement and who has a mastery of even the four fundamental operations. Teachers in the upper grades contend that the pupils who enter their classes are not prepared to carry the work because their thinking is illogical and their ability in computation is poor.

One writer asserts that all the complaints about inaccuracy are an evidence that accuracy is one of the goals which education ought to reach. Ex-President Eliot

43

enumerates among the essentials of the cultured mind the ability to think clear and straight. Accuracy is one of the marks of the scholar. The system of education which minimizes the importance of accuracy of thought and of expression is relegating to a subordinate position one of the essentials of true scholarship and culture.

Accuracy of Thought

If there is one subject rather than another in the curriculum which should be characterized by a high degree of accuracy, that subject is mathematics. In mathematics a statement is either right or it is wrong; there is no middle ground. Mathematical accuracy is proverbial. The demand of the present is for greater accuracy in both thought and computation. Accuracy of thought is the more fundamental and is quite frequently the basis for accuracy of manipulation. Myers defines accuracy of thought in arithmetic as "the degree of closeness of expression to idea—the adequacy of assertion to thought." As the pupil matures, his accuracy of expression should improve, as measured by an absolute standard, for his conception of quantitative relationship is extended. The expression used by the pupil, often exposes the mental process; teachers can improve accuracy of statement by frequently insisting upon full and concise explanation.

A great responsibility rests upon the teachers of the early grades for securing accuracy of expression in oral and in written work. In these grades inaccurate habits of expression are often formed and if not corrected before the pupil reaches the upper grammar grades they are quite likely to persist and to interfere seriously with the pupil's progress in the subject. David Eugene Smith says, "It is the loose manner of writing out solutions, tolerated by many teachers, that gives rise to half the mistakes in reasoning which vitiate the pupil's work," and "teachers are coming to recognize that inaccuracies of statement tend to beget inaccuracy of thought and so should not be tolerated in the schoolroom." There is real value in clear cut and concise statements of conditions and arguments. One who has observed recitations in a large number of schools is impressed by the looseness and carelessness of expression in both oral and written arithmetic. In arithmetic, as in other subjects, it ordinarily takes at least a whole sentence to express a thought.

A number of the more common inaccurate statements found in certain text-books and frequently permitted to go unchallenged in classrooms will now be considered. The fact that certain of the inaccuracies considered are found in text-books of superior merit does not lessen the justness of the criticisms that follow. A statement is not correct merely because it is used by a large number of authors. The question at issue is not whether this or that author has a correct statement for the idea under consideration; we are not considering an author as an individual; "Not who is right, but what is true," is the question at issue. The good teacher will be open-minded and candid with himself and will consider the question in an impersonal manner.

If enough people of influence habitually use an inaccurate statement until its "real" meaning becomes apparent it is hardly worth while to raise objection to the inaccuracy. The inaccuracies which follow are not of this type.

Figures and Numbers

One common inaccuracy is due to a confusion between the symbol and that which the symbol represents. Frequently the pupil is directed by the author or the teacher *to add figures.* Figures are not numbers; they are the symbols which represent numbers. It is just as impossible to add, subtract, multiply, or divide figures as it is to harness the picture of a horse. ·The figure is not the real thing; it is that which represents the concept to the mind. It is possible to perform various operations upon numbers, but these operations cannot be performed upon that which merely represents numbers. It is important that both teachers and pupils discriminate between numbers and the characters which represent them. It is proper to direct the pupil to add the numbers represented. The teacher is reminded in this connection that accuracy is not a little thing, but even in little things we should be accurate.

Inaccuracy in Statement

Such statements as the following are frequently found in classroom work and in some text-books:

$$3 + 4 = 7 + 5 = 12 \times 2 = 24$$

The inaccuracies in the above statement are evident. It is not true that $3 + 4 = 7 + 5$, nor does $7 + 5 = 12 \times 2$. The mere fact that the correct answer is obtained does not justify the inaccuracy of statement. If the data given are correctly used and all the computations are accurate, the answer in any problem will be correct; but it is possible to obtain the correct answer to a problem, even though no step in the solution is correct. This may be brought about by a balancing of inaccuracies in computation. The process may justify the answer, but the answer does not justify the process.

If the above example needs solution, it should be put into the following form: $3 + 4 = 7$; $7 + 5 = 12$; $12 \times 2 = 24$.

Inaccuracy in the Addition of Mixed Numbers

A third inaccuracy is the following:

$4\tfrac{2}{3} = \tfrac{8}{12}$

$3\tfrac{1}{2} = \tfrac{6}{12}$ It is apparent that $4\tfrac{2}{3}$ does not equal $\tfrac{8}{12}$ as the solution asserts, nor does $3\tfrac{1}{2}$ equal $\tfrac{6}{12}$ or $2\tfrac{3}{4}$ equal $\tfrac{9}{12}$.

$2\tfrac{3}{4} = \tfrac{9}{12}$

$9 + \tfrac{23}{12}$; the sum $= 10\tfrac{11}{12}$.

The solution may be correctly expressed in several ways. The following does not violate the truth in any of its statements:

$$4\tfrac{2}{3} = 4\tfrac{8}{12}$$
$$3\tfrac{1}{2} = 3\tfrac{6}{12}$$
$$2\tfrac{3}{4} = 2\tfrac{9}{12}$$
$$\text{Sum} \quad 10\tfrac{11}{12}$$

Inaccuracies in Multiplication and Division

The definitions of multiplication and division are often violated.

In multiplication the multiplier must always be abstract, and the product must always be of the same denomination as the multiplicand.

In division the quotient is always of the same name as the dividend when the divisor is abstract.

Such inaccurate statements as the following are not uncommonly made in classrooms:

$$2 \times \$50 = 100 \qquad \$100 \div 4 = 25$$
$$2 \times 50 = \$100 \qquad 100 \div 4 = \$25$$
$$\$100 \div \$4 = \$25$$

Feet Multiplied by Feet

The following is an inaccurate expression: 4 ft. × 5 ft. = 20 sq. ft. The first factor (4 ft.) cannot represent the multiplier, because it is not abstract; for the same reason the second factor (5 ft.) cannot be the multiplier. Moreover, since the product must be of the same name or denomination as the multiplicand, it is apparent that if the result (20 sq. ft.) is correct, neither 4 ft. nor 5 ft. can be the multiplicand, for they are not of the same name as the product. Since the problem involves neither multiplicand nor multiplier, it is evidently not a problem in multiplication. This multiplying of feet by feet and calling the result square feet is óne of the most common inaccuracies of statement in arithmetic. If a rectangle is 4 ft. wide and 5 ft. long, we may find its area by the following method: $4 \times 5 \times 1$ sq. ft. $= 20$ sq. ft.

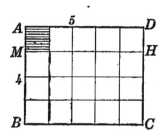

The area represented by ABCD is $4 \times 5 \times 1$ sq. ft. because it contains 4 rows similar to AMHD which contains 5×1 sq. ft.

The inaccuracy of the statement, 4 ft. × 5 ft. = 20 sq. ft., may also be shown by interpreting multiplication as a short method for addition. It is evidently impossible to add 4 ft. (or 5 ft.) to itself any number of times and get a result whose denomination is square feet. It would be possible to extend the definition of multiplication so as to make 4 ft. × 5 ft. = 20 sq. ft. a correct expression, but this has not been done, and there is no necessity for doing so in the elementary school.

Common Fractions

In the analysis of problems in fractions and in percentage, inaccuracies of expression are very common. The following illustration will indicate the nature of these inaccuracies:

Problem. **Two-fifths** of a number equals 12. Find the number.

I	II
AN INCORRECT SOLUTION	A CORRECT SOLUTION
$\frac{2}{5}$ = the number	$\frac{5}{5}$ of the number = the number
$\frac{2}{5}$ = 12	$\frac{2}{5}$ of the number = 12
$\frac{1}{5}$ = $\frac{1}{2}$ of 12 = 6	$\frac{1}{5}$ of the number = $\frac{1}{2}$ of 12 = 6
$\frac{5}{5}$ = 5 × 6 = 30	$\frac{5}{5}$ of the number = 5 × 6 = 30

Not a statement in solution I is true. If $\frac{5}{5}$ equal the number, it must be true that the number equals 1, since $\frac{5}{5}$ equal 1; but the number is 30, and not 1. If $\frac{2}{5}$ equals 12, then $\frac{5}{5}$ must equal $\frac{60}{5}$, the equivalent of 12; but this is evidently absurd. If $\frac{2}{5}$ equals 12, then $\frac{1}{5}$ must equal 6; but 6 equals $\frac{30}{5}$, and $\frac{1}{5}$ does not equal $\frac{30}{5}$. The last statement of solution I is seen to be absurd as soon as we substitute for $\frac{5}{5}$ its equal, 1, for we then have 1 = 30. To say, "Let $\frac{5}{5}$ equal the number," or "assume" that $\frac{5}{5}$ equals the number, does not eliminate the inaccuracy. The incorrectness of solution I is not an incorrectness in reasoning after the first steps are taken. The reasoning is faultless after the first two steps. The incorrectness lies in the inaccuracy of statement. The first two statements are untrue, hence the reasoning based upon them is untrue. Any absurd conclusion may be reached by correctly reasoning from false premises.

At first sight it appears that solution II does not differ greatly from solution I, but a brief study of the two solutions will convince one of the inaccuracy of the first and the correctness of the second. The word "of" when used with common fractions indicates multiplication. Since $\frac{5}{5}$ is a common fraction, the first statement of solution II is equivalent to "*1 times the number equals the number.*"

This is always true. Contrast this with the corresponding step of solution I, which asserts that $\frac{5}{5}$ (or 1) *equals the number*. This last statement is never true, unless the required number is actually 1. The second statement in solution II is given in the original problem, and hence is known to be true. From these two statements, or premises, both of which are true, by accurate reasoning we arrive at the conclusion which is true.

Many teachers may feel that their pupils do not have time when analyzing problems of this type to write an analysis similar to solution II. It need only be urged that no teacher and no pupil should ever be so pressed for time or for space that he hasn't time to tell the truth. It is not correct to say that $\frac{5}{5}$ equals A's age, or A's money, or the distance from B to C. The accurate statements are:

$\frac{5}{5}$ of A's age equals A's age.

$\frac{5}{5}$ of A's money equals A's money.

$\frac{5}{5}$ of distance from B to C equals distance from B to C. It is not intended in the preceding illustrations to convey the idea that the suggested forms are the only correct forms. An accurate statement of facts may often be made in a variety of ways, and it matters little how the fact is stated, provided only that the statement is accurate, concise, clear, and grammatical.

Inaccuracies in Percentage

The analysis of problems in percentage often leads to inaccuracies similar to the one just considered. *"Forty per cent of a number equals 80. Find the number."* If we begin the solution by asserting that 100% equals the number, we are evidently committing the same inaccuracy of statement that was pointed out above, since 100% equals $\frac{100}{100}$, and it is not correct to say that $\frac{100}{100}$ equal the number.

The correct statement is 100% *of the number* equals the number.

Mensuration

It is incorrect to speak of the square root of 49 square feet as 7 feet. By the square root of a number is meant that which multiplied *by itself* produces the given number. Thus, the square root of 25 is 5, because $5 \times 5 = 25$. If the square root of 49 sq. ft. = 7 ft., it follows that 7 ft. × 7 ft. = 49 sq. ft.; but from the argument immediately preceding, it appears that this is impossible. It is incorrect to say 27 cu. ft. ÷ 9 sq. ft. = 3 ft., since 9 sq. ft. × 3 ft. does not equal 27 cu. ft. For justification of this point and for fuller explanation, see the chapter on Mensuration.

Longitude and Time

In the solution of problems in Longitude and Time, pupils should be taught that 15 degrees correspond to 1 hour of time, instead of 15 degrees equal 1 hour of time. The relationship is not one of equality in the mathematical sense of the word.

The preceding inaccuracies are common in our schools, and they are all in direct contradiction to the habits of accuracy of expression which should be cultivated in the study of arithmetic. Moreover, such inaccuracies as those just pointed out are likely to lead to confusion and to inaccuracy of thought. Teachers should not tolerate them in their classes. Such inaccuracies of expression cause "many teachers to sit up at night to correct mistakes which they had better sit up in the daytime to prevent."

Inaccuracy in Computation

Not only should accuracy of statement be secured, but a high degree of accuracy of computation should be acquired. The evidence of inaccurate computation in our schools is overwhelming. It is not unusual to find a group of seniors in high school not twenty-five per cent of whom can add ten numbers of four digits each and secure the correct result the first time.

The business man contends that it is very difficult to secure a boy who can add, subtract, multiply, and divide with speed and with a high degree of accuracy.

When a boy or girl seeks a position in which accuracy and rapidity of computation are essential, it is of little consequence that the applicant can solve problems and explain them in a clear and concise manner, if the ability to perform the arithmetical operations with speed and accuracy has not been developed. In many vocations work that is inaccurate is of no value whatever.

Accuracy in Practical Work

It is not always possible to secure absolute accuracy, but the standard should be set as high as can be attained. The Coast and Geodetic Survey uses an ice-bar apparatus for measuring a base line. An ordinary metal bar would be subjected to expansion, and the results would not be accurate. By supporting the metal measuring-bar in a trough packed in ice, it is maintained at a uniform temperature, and a base line can be measured with an error of only one part in 2.5 million. The surveyor's calculation of areas must be accurate, and so he checks his computations by totals of latitudes and departures. In 1900 the United States Coast and Geodetic Survey completed the measurement of an arc along the 39th parallel from Cape May in New Jersey to Point Arenas in California, a distance of 2,625 miles. So carefully was the work done that the total amount of probable error does not exceed 100 feet.

Many pupils go through the entire school course without appreciating the value of accuracy. "Arithmetic as a tool is almost useless unless it has an edge keen enough to do its work with considerable speed and absolute accuracy. Speed must first be attained by having pupils deal with things so simple that practically all the attention can be given to the speed itself. Teachers permit and sometimes encourage inaccuracy by giving high grades for the correct process, even if the result is absurd. Pupils soon form the habit of accuracy when they find that inaccurate results are always marked zero."[1] The good is the enemy of the best, in the arithmetic class as well as in many other places. It is not easy to secure work that is characterized by a high degree of accuracy. The teacher must continually insist upon accuracy, but the results fully justify the time spent and the effort expended. After a high degree of accuracy has been secured, the speed can be gradually increased, but mere practice will never make the shiftless and careless pupil accurate in his computations. With such a pupil there is almost an inverse relation between speed and accuracy. The faster he works, the greater the number of mistakes he makes. We must not permit our pupils to regard accuracy of computation as a matter of small importance. Some of the pupils of our schools seem to believe that accuracy in computation is of so little moment as to be almost beneath their intellectual dignity. It is hard to eradicate such an idea after the pupil reaches the high school age.

Speed and Accuracy

It was formerly thought that there is a very direct relation between accuracy and speed. It was asserted that the most rapid computers are always the most accurate. The

[1] N. E. A. Proceedings, 1906, p. 101.

relation is not so intimate as was formerly supposed. Stone[1] showed that there is not a necessary relationship between speed and accuracy. In his investigation the school system which ranked first in accuracy was fourth in rapidity, while the system which ranked second in accuracy was tenth in rapidity. The effect of drill on speed and upon accuracy, as indicated by some recent investigations, is considered in detail in the chapter on Drill.

The subject of checks is closely related to that of accuracy of computation. Even the professional mathematician, who makes computations with great frequency, does not guarantee the accuracy of his results until he has applied to them some adequate check. Some simple checks for accuracy of results should be extensively used in the schools, and a number of such checks are considered in the following chapter.

Copying of Figures

Copying figures is now one of the tests in civil service examinations, and because of its importance it should have some consideration in the work in arithmetic. Recently one of the authors investigated a group of seventh grade pupils with respect to inaccuracy in copying figures from a printed page. Mistakes of this type are of two general kinds; first, the inserting of figures that do not occur in the original copy, and, second, the interchange of figures which actually do occur. As an example of the first class of errors, the copying of 59 for 86 might be cited, while the copying of 34 for 43 exemplifies the second class. Errors of the first class were more frequent than those of the second class, in the ratio of 28 to 11.

[1] Stone, ''Arithmetical Abilities and Some Factors Which Determine Them.''

Pupils who are below the average in speed miscopy figures more frequently than those who are above the average. Figures were omitted in the copying about one-third as frequently as they were miscopied.

In a test in the seventh grade involving only addition, subtraction, multiplication, and division, it is probable that from four to eight per cent of all mistakes are due to miscopying or omission of figures. No doubt this percentage is higher in the lower grades.

CHAPTER IV

CHECKS

In the preceding chapter the necessity and the importance of accuracy of thought and expression were considered. The effect of drill upon accuracy of computation is considered in the chapter on Drill. We will now consider some other methods by which increased accuracy of computation may be secured.

Necessity of Using a Check

Long practice in mathematical computations is indispensable to both accuracy and speed, but practice alone is not sufficient to produce the degree of accuracy which is essential in mathematics. Even accountants and professional mathematicians are not willing to guarantee the accuracy of their computations until some appropriate check has been applied. The surveyor checks his long chain of calculation by totals of latitudes and departures; the navigator who computes his position at sea must check his results if he would guard the lives and the cargo entrusted to his care. A mistake of a mile in computing his results may mean the loss of hundreds of lives and of thousands of dollars. If a result will check it is correct, if it will not check it is incorrect. The best check to be applied is determined by a number of factors. The check should vary with the process and with the maturity of the pupil.

Checking by the Teacher

Checks are of many kinds. The teacher may serve as a valuable check upon the work of the pupil, but many teachers, by habitually accepting inaccurate and untidy work, do not stimulate the pupil to greater accuracy. Frequently work that is not correct in every particular should be considered wholly wrong, and the teacher who habitually accepts work that is not accurate is, perhaps unconsciously, discounting the value of accuracy. The teacher should always be a check upon the work of the pupil, but the check of greatest value to the pupil is the one that he himself applies.

Pupil Must Check His Results

Unless a pupil has acquired the habit of checking his results, he is not master of the situation; such artificial devices as answer books are of value chiefly in checking unusually long or involved problems. The earlier the constant verification of results is begun, the more automatically it will be practiced. It is not uncommon for teachers to object to the use of checks on the ground that they diminish the number of problems that it is possible for a pupil to work. The fact that the check itself is a problem is frequently overlooked. The essential thing is to gain "well-grounded self-confidence" to *know* that the result is correct.

The checks discussed in the following pages are used by teachers and by men in the various vocations in which accurate computation plays an important part. Not all of the checks should be taught to any class of pupils. One good check, for a given process, so well known and so frequently used that its application has become almost automatic to the pupil is of more value to him than three or four checks all of which can be used but no one of which has been thoroughly mastered. Many teachers make the mistake of familiarizing their pupils with several checks

but not requiring them to use any one of them long enough
to make it really practical.

Checks for Addition

Checking addition by combining the addends in the
reverse order is so well known as to render comment unnecessary. To combine in the same order the second time
as the first is practically no guarantee of the correctness
of the result. The mind tends to repeat its own mistakes,
and if two numbers are incorrectly combined the first time,
the chances are that the same mistake will be made if the
order of the addends is not changed.

$$
\begin{array}{r}
\text{Add } 4823 \\
6792 \\
8437 \\
9265 \\
7426 \\
\hline
36743
\end{array}
$$

The numbers represented by the first column may be added
and the result is 23. The sum for the second column, without carrying, is 22; for the third column the sum is 25,
and for the fourth it is 34. These sums may be represented
in either of the following manners and then combined:

$$
\begin{array}{cc}
23 & 34 \\
22 & 25 \\
25 & 22 \\
34 & 23 \\
\hline
36743 & 36743
\end{array}
$$

The above methods are frequently used by persons who are
likely to be interrupted while adding.

Casting Out Nines

Another simple and easily remembered check for addition is that of casting out the nines. The following illustration will make clear the method of applying this check. Suppose we wished to check the result, 36743, in the first example cited above. Add the digits in each of the addends and divide the sum by 9. The remainder is called "the excess." (It is the remainder which would be found if the entire number were divided by 9.) After the excess of nines in each addend has been ascertained, add these excesses and divide the result by 9. The remainder (or excess) thus obtained should equal the excess in the sum in the original example.

Excesses

4823	8
6792	6
8437	4
9265	4
7426	1
36743	23 = sum of the excesses.

Excess in the sum (36743) equals 5. This is also the excess in the sum of the excesses.

RULE.—*The excess in the sum must equal the excess in the sum of the excesses.*

Another illustration will be given without explanation:

Excesses

9426	3
3875	5
2873	2
6941	2
2739	3
25854	15 = sum of the excesses.

Excess in the sum (25854) equals 6. This is also the excess in the sum of the excesses.

After the check has been applied several times, the pupil should find it unnecessary to write down the various excesses. After a brief period of drill, the entire check can be applied without writing down any of the figures. Moreover, the pupil will soon discover that it is wise to group the digits of a number into combinations whose sum is 9, and that the grouping enables one to determine the excess in a number very quickly. For illustration, it is desired to find the excess of nines in 32789. The practiced eye will group the 7 and the 2, and will see that the only thing to do thereafter is to determine the excess of nines in $8+3$. Instead of adding the digits and then dividing the sum by 9, pupils should accustom themselves to add the digits until the sum is 9, or more; then add only the excess above 9.

It is apparent that the check by casting out the nines is not a proof of the accuracy of the result. If two digits in the result were interchanged, or if the errors were of such magnitudes as to balance each other, the check would not detect the errors. Moreover, the check is not short, and it is more liable to error than a new addition, unless as much time is given to drill on casting out of nines as to the addition itself, and this is not desirable.

Checks for Subtraction

Subtraction may be checked by casting out the nines, but the well-known checks of adding the difference to the subtrahend to produce the minuend, or of subtracting the difference from the minuend to produce the subtrahend, are more easily applied.

Checks for Multiplication

Multiplication may be checked in a number of ways. Probably the methods most familiar to the majority of teachers are the dividing of the product by the multiplicand to produce the multiplier, or dividing the product by the multiplier to produce the multiplicand. The check by casting out the nines is especially adapted to multiplication, and is easily applied. The following will illustrate the method of applying this check:

Excess

```
4836        3  (excess of nines in 4836)
 285        6  (excess of nines in  285)
─────
24180
38688
9672
───────
1378260    18 = product of the excesses.
```

Excess in the product (1378260) is 0, which is also the excess in product of the excesses.

RULE.—*The excess of nines in the product must equal the excess in the product of the excesses of the multiplicand and the multiplier.*

A second illustration follows:

Excess

```
3725        8  (excess of nines in 3725)
 439        7  (excess of nines in  439)
─────
33525
¹1175
14900
───────
1635275    56 = product of the excesses
```

Excess in product (1635275) is '2' which is also the excess in the product of the excesses.

Next in value to the check by casting out the nines is that of interchanging multiplicand and multiplier. An illustration will make this clear.

427	326
326	427
2562	2282
854	652
1281	1304
139202	139202

Checks for Division

Division may be checked by determining whether the result obtained by multiplying the divisor by the quotient and adding the remainder, if any, to this product, equals the dividend. Casting out the nines may be used to advantage in checking division. The first illustration that follows shows the method of checking division when there is no remainder. The second shows the method when there is a remainder.

Divide 1635275 by 439.

```
439)1635275(3725
    1317
    ────
    3182
    3073
    ────
    1097
     878
    ────
    2195
    2195
```

The excess in the dividend (1635275) is **2.**
The excess in the divisor (439) is **7.**
The excess in the quotient (3725) is **8.**

The product of the excesses in the divisor and quotient is 56, and the excess in 56 is 2. This is also the excess in the dividend.

RULE.—*When there is no remainder, the excess in the dividend equals the excess in the product of the excesses of divisor and quotient.*

When there is a remainder, the check is applied as follows:

$$847)57426(67$$
$$5082$$
$$\overline{6606}$$
$$5929$$
$$\overline{677}$$

Excess in the dividend is 6; in the divisor it is 1; in the quotient it is 4; and in the remainder, 2.

The product of the excesses in the divisor and quotient (1×4) is 4. Add the excess (2) in the remainder; the sum is 6. (Cast the nines out of this sum if it is greater than 9.) The result, if correct, must equal the excess in the dividend.

When one's attention is first directed to these checks they seem quite long and involved, but practice will enable one to use them with facility.

Approximate Checks

A rough or approximate estimate of the result before a computation is made is frequently a valuable check. Too often pupils submit an answer of $480 when the result should have been $4.80, or a result of $54.16 when the correct result is a hundred times as large. If teachers would drill their pupils more frequently in approximating results, the number of absurd and impossible answers would be reduced. If a problem requires the cost of 520 yards of cloth at 13 cents a yard, the pupil should know whether the result is about $6.50, $65, $650, or $6500. If the problem requires the cost of 432 articles at 34 cents each, the pupil should see that since the cost is about $⅓ each, the entire cost will be about one-third of $432, or $144.

Self-Checking Problems

Self-checking examples and problems serve a useful purpose, but checking should not be confined to this type. The following will illustrate this type:

$$3+ 5+ 7+ 6=21$$
$$2+ 8+ 9+ 3=22$$
$$5+ 4+ 8+ 9=26$$
$$7+ 2+ 6+ 5=20$$
$$\overline{17+19+30+23=89}$$

The admissions to a certain ball park were as follows:

	Wed.	Thurs.	Fri.	Sat.	Totals
Children............	428	369	514	483·	1794
Women.............	387	563	618	327	1895
Men...............	942	1048	1127	1026	4143
	1757	1980	2259	1836	7832

Numerous self-checking problems based upon local conditions will suggest themselves to most teachers. The election returns, the enrollment in various grades of the school, the number of pieces of mail of various classes handled by the local postoffice in a month, are among the problems that may be used.

Checking Problems

Reasoning problems may often be checked by solving in two ways and comparing the answers. Instead of using unitary analysis, proportion may be used; instead of solving a problem involving cost in denominate numbers by the use of aliquot parts, cancellation may be used.

Occasional checking is of only slight value. The pupil should get correct results and should know that the results are correct. Habitual checking of results tends to beget confidence in one's results, and confidence in one's ability to produce correct results in a given field lies near the basis of success in that field. The pupil who lacks assurance in himself may actually retrograde in his work because lack of confidence undermines his power of action.

CHAPTER V

MARKING PAPERS IN ARITHMETIC

Objections to Marks

No system for marking papers has been devised that is free from objectionable features. Numerical marks are generally used to represent the value or the correctness of the result, but they usually put no value upon the effort required to obtain the result. Children are prone to compare one grade with another without taking into consideration individual differences in ability. When grades are given upon the quantity of work done in a given time, the gifted child will, according to our present system, receive a higher grade than the dull child, but the grade of seventy or eighty that the dull child receives may represent highly creditable work for him.

Teachers generally complain because of the amount of labor required in marking papers. This labor may be unprofitable and even harmful to the teacher, but it is one of the necessary burdens of successful teaching. The examination of papers should give the teacher an index to the attainment of the pupils and, perhaps, indirectly, a measure of her efficiency in teaching, but the marks and the criticisms given should serve as a certain stimulus and incentive to the pupils to work. The marks, checks and criticisms given should not be looked upon as mere awards for effort or achievement but as suggestions for improvement. Wherever they do not serve as educational stimuli, it is questionable if they should be used.

A Daily Memorandum

Some marking is necessary for mere bookkeeping purposes. With large classes it is more or less impossible for the teacher to remember the difficulties and needs of each individual of the class. Instead of assigning grades at the close of the recitation, the teacher will find it more practicable and helpful if she keeps a memorandum in a book, recording the entries for each child upon a separate page, as follows:

CHARLES GORDON

Oct. 3. Seems uninterested, but understands the principle.

Oct. 4. Work may be uninteresting because it is too easy. Give special problems.

Oct. 12. Some improvement. Problems may be too abstract. Try more concrete ones.

Oct. 14. Likes to make his own problems, but doesn't know prices well enough to make his problems harmonize with business procedure.

REX SAFFER

Oct. 5. Lacks facility in fundamentals.

Oct. 7. Assigned problems for outside practice.

Oct. 8. He reported the work done. Gave him a chance to show how quickly he could do them. Improvement. Commended him.

Oct. 9. Assigned more problems for outside practice.

Oct. 12. Assignment worked. Takes pride in it. Continue the practice.

Such records as these would be of great value to the teacher at the close of the term (in determining the grades of students).

How to Get the Best Results from Marks

In order to get the best results from a written exercise, the papers should always be marked and handed back to the pupils. Ordinarily a teacher should mark but not correct the mistakes. That is the pupil's duty. However, the notes of the teacher should be clear enough to enable the pupils to correct the errors intelligently. It is well to remember that calling attention to a fault or a mistake once does not mean that it will be corrected. Attention must be called to it repeatedly until the pupil habitually gives the right form. A clear distinction should be made between errors due to ignorance and those due to carelessness. Again, not all the faults can be corrected at once. The most serious mistakes should be overcome first; the minor ones should be taken up later. Perfection is attained by gradual growth.

Stone's Plan

In preparing his study in Arithmetical Abilities, Dr. Stone used the original scheme, in marking the papers, of giving a score for each step in a problem. This is an especially good scheme to employ in grading the fundamentals in the primary and intermediate grades. Of course such a system of grading is based upon the assumption that one combination is of equal difficulty with another. In such a plan a student gets credit for work done correctly. If two of the three columns are added correctly, the student gets two credits, although the answer as a whole is wrong. This plan could not be applied to problems unless they were of equal difficulty or were graded according to difficulty. At present we have no scale of problems for measuring arithmetical ability. Such a scale will surely be forthcoming in the near future.

Lack of Uniformity Among Teachers as to Marks

The authors attempted to find out from teachers what the current practice is in regard to the marking of papers in arithmetic, but they found no common mode of procedure. Some teachers take into consideration spelling and punctuation; others, the minuteness of analysis; others, the general form of the paper; and still others, the character of the handwriting. Although the solution may be mathematically correct, it is an easy matter to find teachers who refuse to give full credit for it, if there is a misspelled word in the solution or if an interrogation point is omitted. At times this matter of form should be the thing of paramount importance, but to place great emphasis upon it in every written exercise means that it was not well taught at the proper time. In written exercises the good is frequently the enemy of the best. The fact that the first copy should be good enough to keep should be impressed upon the pupils.

Marking the Habit and Reasoning Phases of Arithmetic

In the habit phases of arithmetic there are only two things to be tested: accuracy and rapidity. If the various combinations are to be made automatic, rigid tests must be made repeatedly to avoid wasteful practices and to insure progress. It is quite as necessary to note the kind and number of mistakes made as to note the number of combinations correctly made. In fact the number of mistakes and the frequency with which given combinations are erroneously stated, is the very best measure of accuracy. If progress is being made, the tests given by the teacher will show a decrease in the number of mechanical errors.

After the combinations have been mastered, the pupils should be drilled and tested for rapidity. The best results will usually be secured if the pupils compete against time. With watch in hand the teacher directs the class to write certain multiplication tables or to do certain sums against time. For this to be of the greatest value the time records of each individual and of the class should be preserved for future comparisons. The amount of time necessary for any given type of work should decrease rapidly at first and more gradually as the pupils approach their psychological limits.

In grading problems in arithmetic there is only one new element to be added,—the number of steps in the problem. If these are understood in their sequence, the rest of the work is a mere matter of computation. Until we have a more scientific basis for estimating these steps, they may be regarded as of equal value or of equal difficulty. In grading problems which require interpretation credit may properly be given both for interpretation and for computation.

It is usually impossible to grade work of any kind to within one percent. The marks of excellent, very good, fair, not satisfactory, etc., usually serve the purpose better.

CHAPTER VI

THE NATURE OF PROBLEMS

The Changed Character of Problems

Text-book makers are always limited in the selection and organization of their material by two facts: first, changing social conditions, and, second, the strengths and limitations of the different maturity levels of childhood. The text-book and the text-book makers are, therefore, intermediating agencies. They are confronted with the problem of making a double adjustment; inwardly to the human nature of childhood, and outwardly to those social forces and conditions that largely determine the nature and organization of the school.

New materials come into the curriculum in response to new needs, or in response to old needs spreading over wider areas, or whenever there is a redistribution of the burdens resting upon established institutions. Tradition and reason have been unable to set a fixed division of work upon social institutions. As society becomes increasingly conscious of the value of incidental agencies of informal education of one century, it frequently gathers them up and imposes them upon the schools. For this reason the new arithmetics contain numerous problems concerning the saving and loaning of money, mortgages, improved banking, building and loans associations, bonds, rentals, taxes, public expenditures, and insurance. Moreover, there is an increasing tendency for such problems to be arranged and presented in the grammar grades to correspond to the occupations they illustrate. Thus we have a list of problems dealing with agriculture, another with banking, another with household science, and another with trade conditions. This is one of the numerous attempts to adjust the school to vocational life.

It is also held that material thus arranged gives the student an abundance of valuable and interesting information, and that it does not in any sense diminish his opportunities for acquiring arithmetical skill. Arithmetic is not taught primarily to give information about the Keokuk Dam or the Panama Canal, but in understanding and comprehending the Keokuk Dam and the Panama Canal some arithmetic is necessary. The solution of occupational problems does not rationalize arithmetic; such problems merely show a correlation of material in one field of thought with outside events, forces, and conditions. Involved in this attempt to emphasize the social and industrial phases of arithmetic are two highly desirable outcomes, (1) a knowledge of a more extended application and use of arithmetical forms and skills to the work of the world, and (2) their value and use in aiding us to interpret our daily experiences.

Writers of arithmetics make little effort to localize materials. This is clearly the province of the teacher. It is easily possible for teachers to carry the practice to an unnatural extreme. The authors knew of a teacher who lived in a community that gave considerable attention to poultry raising. Under the direction of this teacher the pupils devoted one entire term to the preparation of original problems dealing with poultry raising. Such extravagant attempts at local adjustment as this are regarded as ridiculous by the conservative schoolmaster.

Abstract and Concrete Problems

Problems are most commonly classified as abstract or concrete. There is a tendency to increase the number of concrete problems. In the lower grades these problems are selected from play, school, and home activities. The attempt is made to introduce no problems that are more difficult or complicated than those the children use in their daily life outside of school. In this there is a well pronounced danger—that of keeping children upon a low level of arithmetical ability until outside necessity forces them to acquire new skills. The problems used in school should frequently be slightly more difficult than those met in outside life—should require some "stretching up." Otherwise the school is not fulfilling its purpose as a time and labor saving device.

On the other hand, some have an erroneous notion as to what constitutes a concrete problem. Whenever simple abstract numbers are labeled with some concrete denomination, like dollars, pounds, or acres, it is assumed that the problem has been made concrete. Such an assumption may be far from the fact. No matter how concrete the labels may be, unless the facts have a basis in the child's experience, the problem is abstract to him.

There is still a third way in which the tendency to make arithmetic more concrete may lead to harmful, if not wasteful, results. It is the practice of labeling all the numbers in the processes of calculation. To illustrate:

1. Chester had 13 marbles and his father bought him 15 more; how many had he then?

2. A man bought 7 head of cattle at $45 a head; what did they cost him?

3. If $56 are divided equally among a number of boys, so that each boy receives $7, how many boys are there?

4. If 30 oranges are divided equally among 5 boys, how many will each get?

(1)	(2)	(3)	(4)
13 marbles	$ 45	$7)$56	5 times) 30 oranges
15 marbles	7 times	8 times	6 oranges
28 marbles	$315		

Referring to the above solutions, Dr. Faught says: "The impression seems to prevail that the processes of calculation are in this way made concrete to the child. These problems can be solved concretely; but this would require us to count the marbles, or divide the oranges among the boys. The processes of calculation are intended to avoid this method of solution; they have their only explanation in the fact that they enable us to obtain certain numerical results more economically and quickly than by dealing with the actual things themselves. But is there no place for the concrete in the solution of problems? To be sure, the child must be led to see the process or processes implied in the problem by appealing to the concrete, to objects, drawings, etc., and even to the actual things themselves, if necessary. That is the vital point in the solution of a problem, and the one at which so many teachers fail. But after the process or processes have been determined, it is a question of the abstract, of symbols."

"The word marbles does not help the child in thinking that 5 and 3 are 8, or the $ in thinking that 8 sevens are 56. Neither does he, nor should he, have a mental image of marbles or dollars in thinking these relations. To insist on the boy counting his marbles after each number in the first solution is just as foolish as to insist on the boy counting his marbles thus: one marble, two marbles, three marbles, four marbles, etc. Boys do not count their marbles that way."

The method described in this quotation is the one children and adults use. They use the processes of calculation in the shortest possible way for finding numerical results. They never stop to count the things; they seldom know or try to remember such principles as "If the dividend and divisor are both concrete the quotient is abstract," or "The product is always of the same denomination as the multiplicand." Labels certainly do not assist in calculations, and it is doubtful if they aid materially in making the process concrete. It is no more difficult to think the $45 as abstract in the multiplication than it is to think of cattle as abstract.

Oral and Written Problems

Of course all problems are mental problems. There can be no other kind. "Mental arithmetic" is a misnomer. Usage employs it to mean "oral arithmetic." Racially oral arithmetic must have preceded written arithmetic. Necessity invented the abacus to take the place of counters, pebbles, and fingers. After writing was invented and written problems were introduced a long struggle began, yet unsettled, as to the proper distribution between the two types of problems. So long as the school remains sensitive to social changes this will remain an unsolved problem. No perfectly satisfactory adjustment could be made unless we were to establish a monotonous plane of life. Change of emphasis is inevitable. The number of oral problems has greatly increased in recent years, perhaps relatively more rapidly than the number of written problems. A person is considered illiterate, or practically so, if he must resort to paper and pencil to solve the simple problems of his daily life. A moment's reflection shows how greatly oral problems have increased. Every store; the grocery, the bakery, the hardware, the clothing, and general merchandise stores, sell numerous articles, formerly considered as luxuries, that are now commonly regarded as necessities. Consequently, the average person is compelled to solve a greater variety of oral problems than formerly. No doubt as society grows more complex and varied in nature the number of such problems will continue to increase. Writers of text-books in recognition of this new demand now give long lists of oral problems drawn from daily life. In one of the newer sets of arithmetics 1800 such problems are given.

In this connection it is proper to correct the impression of those teachers who think that rapid work on written problems implies a corresponding degree of speed and accuracy in oral problems. Reliable investigations show conclusively that a person may be rapid in written work and slow in oral work. It follows as a matter of course that many oral problems are necessary if facility and skill are expected in the purely mental operations.

We have alluded to the form of solution of written problems, and have urged teachers to avoid the superfluous exercises not practiced by business men in getting the "answer." We should be seriously at fault and open to criticism if we failed to urge the necessity of rationalizing and explaining most of the mechanical operations and of avoiding inaccuracies of statement. Analysis by steps is needed to familiarize children with the reasoning processes involved in the solution of the various types of problems. The virtue of an analysis is not in its length, but in its directness. Direct analysis means a greater reliance on oral work.

Solution of Problems

It is frequently good practice to separate the solution from the operation. Pupils may indicate the separate steps they would take in solving the problem before they attempt to do the work. Then the solution may be stated free of labels, for example:

1. William had 7 marbles; his sister gave him 4 more, and he traded 3 off for a top; he then found 14 and lost half of what he then had? How many had he left?

SOLUTION:

$$\frac{7+4-3+14}{2}=11$$

11 equal the number of marbles he had left.

2. A field is 80 rods long by 150 rods wide. How many dollars is it worth at $20 an acre?

SOLUTION:

$$\frac{80\times150\times20}{160}=1500,$$

which is the number of dollars this land is worth.

The practice of having pupils indicate the operations should be continued in the upper grades. Whenever the mode of procedure is clear and the habit is fixed for a given type of problem, the practice may be discontinued. Such a device discourages the "trial and error" method normally used by children.

Examples and Problems

The word "*problem*" is used in some sections of the country to mean a statement from which the pupil must first decide what operations are to be performed and what their sequence should be before proceeding to the calculations. In an *example* the mathematical symbols tell the pupil what to do, whether to add, subtract, multiply, or divide.

"An example is a prereasoned problem." Chronologically problems must have preceded examples. The race was confronted with a multitude of interesting problems long before it invented a mathematical symbolism to express number relations. Examples thus became devices for fixing the identical and constantly recurring elements in problems. In spite of this obvious relationship, they are frequently sharply separated and taught as if neither had much dependence upon the other.

Logically, problems should both precede and follow examples in instruction. The formal phases may be abstracted for memorization, when pupils are reasonably familiar with them in their problematic setting, but they should be restored to problems later to insure their application.

Examples are abstract; problems are concrete. Examples are expressed by signs; problems by descriptive statements. In a problem the pupil has the double duty of reasoning and manipulating in either of which he may err. Examples are thus mechanical, problems are rational. An example may be an illustration of some problem or principle that has been demonstrated; a problem requires solution. Examples help to fix things; problems, to insure growth.

Problems Without Numbers

Problems without numbers develop clear imaging and accurate reasoning. The pupil must concentrate upon the reasoning processes, as there are no numbers to confuse or mislead. Such problems tend to develop the ability to decide what operation to use under given conditions. There is a growing demand for more problems of this kind. Any resourceful teacher can manufacture an abundance of them. We give a few illustrative examples:

If you know the height of a flagstaff on which there is a mark, and how far it is from the ground to this mark, how can you find how far it is from the mark to the top of the flagstaff?

To find the area of a square, what must you know and what must you do?

If you know the dimensions of a square, how can you find its perimeter?

If you know the area of a triangle and the length of the base, what else must you know and what must you do to find the altitude of the triangle?

If you know the perimeter and two sides of any triangle, how can you find the remaining side?

' I know how many feet long and wide a cellar is to be, what else must I know and what must I do to find how many wagon loads of earth will be taken out in making the excavation?

I have some money, I know the cost of one hat; how am I to find how many hats I can buy at the same rate with the whole of my money?

B owns a triangular lot; he knows the length, in rods, of its base, altitude, and hypotenuse. How shall he find how many acres it contains?

John weighed a basket of wood; after it stood out in the rain over night he weighed it again. How can he find the weight of water absorbed?

H holds a certain number of books in his hand and tells W that they are one-half per cent of all his books. How can W find how many books H has?

A and B are a certain number of miles apart and are traveling in the same direction. We know the number of miles per hour that each travels and that B travels the faster. How shall we find the number of hours required for B to overtake A?

I know the length of a line of fence, and the number of posts, counting the end ones. How shall I find the distance from one post to the next?

What must you know and what must you do to find how many times a wagon wheel will turn in going two miles?

If you know the length of the edge of a cube, what can you find and how would you do it?

A cow is tied to the corner post of a square lot. Given the length of the rope, how can you find the area of ground over which she can graze if she is inside the lot? If she is outside the lot?

If you know two-thirds of a certain number, how can you find the number?

If you know how many quarts of milk a family uses each day and the price per quart, how can you find the amount of the milk bill for the month of January?

If you know the cost of a bushel of anything and wish to know the cost of a peck, at the same rate, how do you proceed?

What must you know and what must you do to find how many bushels of wheat a bin will hold?

If you know the depth of rain-fall on a given field for a given time, what else must you know and what must you do to find the weight of the water?

What must you know and what must you do to find out how many bunches of shingles would be used in putting a new roof on your house?

If you know how much a man's average expenses are for a month what must you know and what must you do to find out how much he saves in one year?

What must you know and what must you do to find the balance in Mr. A's bank account at the end of a given month?

If you know the cost of a ticket from your home to New

York City what else must you know and what must you do to find the average cost of the trip per mile?

What must you know and what must you do to find the cost of laying a cement sidewalk in front of your home?

What facts must be known and how can they be used in order to compute the average attendance at your school for the year?

Such work as has been suggested above gives the pupils a better appreciation of what the various arithmetical processes do than they otherwise have. It emphasizes the importance of knowing exactly what is to be done before attempting to do it. It makes its first appeal to the understanding rather than to the memory.

Original Problems

Children should be permitted to "make up problems." This is another excellent device for stimulating vigorous thinking. The best of the original problems might be written in a book called "Our Original Arithmetic" or "Our Own Arithmetic." The problems thus preserved may be used later for review purposes. Whenever the data involved in such problems are of an informational character they should correspond to actual conditions. For example it is not wise to assume in a problem that the distance from New York to Chicago is 250 miles; that wheat sells for $6 a bushel and silk for $85 a yard. The duplication of a difficult problem by a simpler original problem frequently clears away the difficulty.

NOTE—For good selections of problems similar to the above the reader is referred to "Problems Without Figures," by S. Y. Gillan; published by S. Y. Gillan and Company, Milwaukee, Wis.

CHAPTER VII

RULES AND ANALYSES

Not many decades ago a teacher who could perform long
and involved computations and could set them down in
neat form was considered a master of arithmetic. The rote
system of learning was in vogue and more emphasis was
placed upon forms and symbols than upon the things
symbolized. Herbert Spencer said that to repeat the
words correctly was everything, to understand the meaning was nothing, and thus the spirit was sacrificed to the
letter. Text-books in arithmetic contained numerous definitions and rules which the pupils were expected first to
memorize and then apply. As an aid to the memory many
definitions and rules were put into rhyme. Rhyming arithmetics became very common during the seventeenth century. The following from the "Handmaid of Arithmetic
Refined," published by Nicholas Hunt in 1633, will
illustrate:

"We are taught in numeration, number riting and notation.

Add thou upright, reserving every ten,
And rite the digits down all with the pen.

Subtract the lesser from the great, noting the rest,
Or ten to borrow you are ever prest,
To pay what borrowed was think it no paine,
But honesty redounding to you gaine."

It is not necessary, however, to go farther back than the
nineteenth century to find examples of these rhyming
arithmetics. A book compiled by John Graham (an Ameri-

82

can) in 1824, entitled, "The Farmer's and Mechanic's Assistant and Companion; or, a New System of Decimal Arithmetic, adapted for the easy and regular instruction of the youth in the United States," was written partly in verse. The author says: "I have endeavored, for the encouragement of the learner, to do all that I possibly could; the rules are all very plain and easy to be understood to which I would advise every scholar particularly to attend; for I, myself, have observed,

> "Study the rule, your question pry,
> You'll gain the answer by and by."

A little further on he says:

> "This little book, peruse it well; I hope in it you'll find
> Something entertaining, useful work, to cultivate the mind;
> From it you may, if well applied, some information gain;
> Arithmetical rules you'll find, both easy, short and plain,
> My best advice to youth I give improve your golden span;
> Seek for knowledge while you're young—education makes the man."

Here follows one of the most interesting problems.

> "Dear friend, I request you with caution and care,
> To measure this lion exact to the hair,
> His head is twelve inches, from his ears to his nose;
> This measure I give you the rest to disclose.

His tail is as long as his head, and a half
Of his body is the length of his head and his tail;
He's a surly old rogue, yet you can not well fail
To tell his whole length, when his substance you see;
You've the length of his head, as was given to me.
The question required, is separately to tell
His whole length, his body's length, and the length of his tail.''

Rules were too often, both the beginning and the end of teaching.[1] It is true that efficiency in a subject like arithmetic presupposes that many processes can be used mechanically, but it should not be inferred that our knowledge of all processes in arithmetic should be mechanical.

We should both educate and train. DeMorgan said, ''The merely showing the student a rule by which he is to work, and comparing his answer with a key to the book is not teaching arithmetic any more than presenting him with a grammar and a dictionary is teaching him Latin.''[2]

Rationalizing the Processes

One of the tendencies in the teaching of arithmetic to-day is the attempt to rationalize the various processes. Every rule in arithmetic rests upon some principle and a topic should be introduced, in the later grammar grades especially, in such a way as to reveal this principle. It is unquestionably true that the rationalization of some processes is difficult for pupils at the age when these processes should be learned. A process may properly be rationalized whenever a pupil of ordinary ability, by whom it is to be mastered, can comprehend it. Some pupils will profit but little by this attempt at rationalization, but this does not justify the teacher in omitting the explanation and

[1] Spencer said that he doubted if one boy in five hundred ever heard the explanation of a rule in arithmetic.
[2] DeMorgan, ''Studies in Mathematics,'' p. 22.

presenting the subject in a mechanical way. Many children can early be trained to look for reasons and laws. "The pupil is expected in a sense to rediscover the subject, though not without profit from the fact that the race has already discovered it. The pupil is a child tottering across the room, not a Stanley penetrating into the heart of Africa. The teacher stands before him and with word and smile entices him on; selecting his path, choosing every spot where he is to plant his foot, catching him when he stumbles, raising him when he falls, but when he has crossed the room he has done it himself and has made more progress towards walking whither he would than if he had been carried across the room, or across hundreds of rooms, or even into the heart of Africa." [1]

Telling is frequently not teaching. The phrase "go and see" has been revised to read "see and think."

No one doubts that children can think and do think about matters that are within their capacities. All normal pupils have a natural desire for knowledge and this desire may be increased and intensified. It is one of the duties of the teacher to make the pupil conscious of his powers; this cannot be accomplished by continually feeding him on an easy intellectual diet. Colonel Parker said that many naturally capable children are "helped into helplessness." The teacher who merely stores the pupil's mind with facts has done only a part of his duty. The pupil should be so trained that he will have the desire to acquire more knowledge for himself and to use this knowledge in an efficient manner. The teacher should seek to make the pupil in the later grammar grades dissatisfied with knowing only the *what* and the *how,* and should emphasize the *why.*

We recognize to-day that the thought side of arithmetic should receive greater emphasis than it did a few genera-

[1] "The Teaching of Mathematics," pp. 69-70. J. W. A. Young.

tions ago. We are breaking away. from the rules of the subject and are directing more attention to the principles that underlie the rules. Arbitrary rules committed to memory cannot develop a thinker.

It is much easier to say to the pupils,—learn this rule and work this list of problems by it,—than it is to teach him to think independently. The indolent teacher will usually choose the path of least resistance and will teach arithmetic in a mechanical way.

Too many pupils in arithmetic are short-circuited and taught results only. In many schools pupils are allowed to combine the numbers in such a way as to get the result and nothing more is expected or required. No one questions the desirability of getting correct results, but when ''getting the result'' is considered as of prime importance and the principles underlying the processes are regarded as of little importance we tend to develop the art of ciphering at the expense of the reasoning faculty. It is not enough simply to do in mathematics. It is important to know also when a process should be used and to understand the reasoning upon which the process is based.

If the principles underlying the various processes are understood and a fair mastery of the fundamental operations is acquired in the first four or five grades, the result will in many cases take care of itself. When the higher purposes are served the lower ones are usually taken care of also. Instead of requiring pupils in the later grammar grades to learn numerous arbitrary rules let us explain the various processes to them, showing how each step is based on something that precedes and how it is to be used in that which follows. These are matters that are worth while. Any method of teaching that thwarts the natural movement of the mind is not efficient. Understanding should usually come before practice. The reversal of this order seldom leads to accumulation of facts or the development of power.

Place of Rules in Arithmetic

Rules have a very important place in the modern teaching of arithmetic. A rule should come at the close and not at the beginning of a process. The rule should be reached and formulated by the pupil himself under the direction of the teacher. It should be a brief and concise statement of procedure for future guidance. The pupil who appreciates the full significance of the favorite aphorism of Lucretia Mott—"Truth for authority and not authority for truth," as it applies to arithmetic, has the proper insight into the subject.

The pupil who has been taught his arithmetic by memorizing dogmatically stated rules sees but little unity in the subject. He is prone to depend upon typical cases and to associate a process with certain phraseology. The pupil who has learned his arithmetic by mastering the principles of the subject is not in serious difficulty every time he encounters a new type of problem, and an insignificant change in phraseology does not mislead him. Any knowledge which the pupil has acquired himself becomes more thoroughly his own than it could otherwise be. "While rules lying isolated in the mind—not joined to its other contents as outgrowths from them—are continually forgotten, the principles which these rules express piecemeal become, when once reached by the understanding, enduring possessions. While the rule-taught youth is at sea when beyond his rules, the youth instructed in principles solves a new case as readily as an old one. Between a mind of rules and a mind of principles, there exists a difference such as that between a confused heap of materials and the same materials organized into a complete whole with all its parts bound together."[1]

Analysis

If it is desirable to emphasize the reasons underlying the various processes, making each step rational to the pupil, it is necessary that some power of analysis should be developed. The power to analyze and to relate is one of the essentials of the clear thinker. The difficulty in a process or in a problem usually lies in the fact that the pupil lacks the ability to break it up into its several parts and to apprehend the relation between the parts. Unless a pupil has the power to analyze, his knowledge of arithmetic will be more or less mechanical. He will depend upon type forms and similar cases and when these cannot be found he is involved in difficulty. A pupil who has the power to analyze is much less dependent upon rules and type forms. He is frequently able to devise a solution of his own and he relies less upon teacher and text. The ability to discover relations and to make proper comparisons gives one relative freedom from mechanical and initiative procedures.

Essentials of an Analysis

An analysis is an orderly statement of the facts; if the facts are understood and their sequence is apprehended the analysis can be stated. Every step in a good analysis is a judgment and is related directly or indirectly to every other step. An analysis should be so concise that if a step is omitted no further progress can be made. Analysis in arithmetic may be over-emphasized, but this should not

[1] Herbert Spencer, "Education," p. 103.

condemn analysis; it should argue for more wisdom in its use. Excess is always to be avoided. No thoughtful teacher would require a pupil to analyze every problem. To do this would be almost as great a mistake as to require no analysis. The correct solution of a problem is usually evidence that a proper analysis has been made. It is possible to train pupils to analyze minutely and yet these same pupils may have but little grasp of number relations. It is not desirable to emphasize either mechanical computation or analysis at the expense of the other. Both are very important and should receive due consideration. Pupils should frequently be required to solve problems by the shortest possible method with no formal analysis or explanation. At such times rapidity and accuracy of solution should be regarded as of paramount importance.

The pupil who has the power to analyze minutely but who performs the mechanical operations slowly and inaccurately is as truly to be pitied as the one who performs these operations with speed and accuracy but has little ability in seeing relationships.

Analyses are sometimes necessary to reveal the line of the pupil's reasoning to the teacher. It is well at times to have the pupils' mental processes exposed. Such a procedure frequently tends to clarify difficulties in the mind of the pupil. Many problems which seem very difficult and involved are easily solved when a detailed analysis is attempted.

Type Forms of Analysis

Some teachers believe that a certain type form of analysis should always be used in the solution of a given type of problem. They insist that the pupil shall always use the exact words of a model analysis which has been explained. It is a mistake to insist upon a rigid and inflexible form in the solution of any problem. There is no type form that is best for all pupils or for all problems. Too great insistence upon particular ways of doing things tends to check originality and initiative. An analysis should be considered as a means to an end not an end in itself. It is not unwise to direct the attention of the pupils occasionally to certain type analyses. Such models may serve as a good basis for others, but there should be no insistence upon the adoption of a particular phraseology. Whatever form most concisely expresses the ideas involved should be used and flexibility of thought and of expression should be encouraged.

Too many words often betray a poverty of mind or lack of a clear comprehension of the data involved. A multiplicity of words may obscure the thought. "It is of little importance how the pupil begins or how he ends the analysis, or whether he puts in the requisite number of 'sinces' and 'therefores' if only he has been,—direct, clear, concise, coherent, and grammatical."[1]

Unitary Analysis

The unitary analysis is of value, but to require the pupil to use it in the solution of every problem is a great mistake. If a pupil really sees the relations between the magnitudes involved he will not often need the unitary analysis. Such an analysis is seldom simpler than a solution by means of a direct comparison. If a pupil is required to solve a problem like the following: *"If 14 spools of thread cost 70 cents, what will 42 spools of thread cost at the same rate?"* he should see at once that forty-two spools will cost three times as much as fourteen spools, and he should be encouraged to use this short cut

[1] Longan's "First Lessons in Arithmetic," p. 8.

to secure the result. In solving such a problem there is a loss of time in first finding the cost of one spool. If the problem stated the cost of 14 spools and required the cost of 33 spools at the same rate, no time would be lost by first finding the cost of one spool. If the cost of twenty articles is given and the pupil is required to find the cost of ten articles at the same rate he should see at once that the required cost is one-half of the stated cost.

Mechanical Analysis

Most teachers have heard of the pupil who when called upon to give an analysis for a certain problem arose but was silent. The teacher said, "Don't you know how to analyze that problem? It is the kind that begins with '*since.*'" With this cue the pupil at once went through a so-called analysis and the teacher commended him for his work. Such a procedure should not be dignified by the name of "analysis." The pupil was only repeating a memorized type form. The analysis had little or no content in his mind. He didn't understand the gist of it. An advertisement frequently seen to-day contains a statement that is in point in this connection, "You may teach a parrot to say 'just as good,' but he won't know what he is talking about." Many analyses which seem to involve a good deal of thought are repeated mechanically when the proper cue is given.

Encourage the pupil to seek for the briefest and best form of analysis. By the use of judicious questions make sure that the analysis has the proper content in his mind. Encourage originality in solution and flexibility of expression.

CHAPTER VIII

THE VALUE OF DRILL

The educational pendulum always oscillates between extremes. Those golden means which the practical administrator desires are seldom discovered. Much of the arithmetic of the past was devoted to training the memory. Now under the influence of the movement for the training of the higher rational processes we are in great danger of failing to reduce to an automatic basis the skills formerly emphasized. This may account partly for the criticism that practical men urge against the product of the schools. Those earlier modes of instruction that called for an automatic mastery of the fundamentals should not be discarded without a hearing. If social demands are to be taken as a criterion for judging the value of the various subjects, then the school should vigorously insist upon a mastery of the materials and processes of arithmetic. This subject, so far as it relates to common community life, increases in serviceableness in proportion to the degree to which its fundamental processes have been reduced to the plane of habit.

Arithmetic: A Habit Study

Arithmetic is not primarily an informational subject; it is primarily a habit subject. By this statement we do not mean that one acquires no information in the study of arithmetic, but that the acquisition of facts should be a secondary consideration. Information is an essential outcome of the study of arithmetic, but the habits acquired are elemental or fundamental. We would not eliminate from arithmetic problems that convey information about business life; in fact, we would multiply them. But in doing so we would not neglect to give a proper emphasis to the other aspects of the subject.

Formal Versus Rational Drill

There are two current notions as to the manner in which habits in arithmetic should be formed,—one old and the other new. There are those who assert that children should become expert in handling the tables before they put the combinations to use in solving problems. About twenty-five years ago reading was taught by the alphabetic method. After the children had learned to repeat the alphabet forward and backward and could combine letters into mono-syllabic words, the teacher permitted them to read a few sentences. Music was begun by singing the scale; the children were told that if they learned it perfectly some day they would be permitted to sing some songs. In drawing, the children first learned how to make horizontal, vertical and oblique lines; these were later combined into geometric figures. In other words the method of instruction accepted and used a quarter of a century ago, was that the habit phases of each subject should be taught and mastered without reference to their use. The illusive hope was held out that mastery was essential because the date, the place, the note, or the number combination, might be needed. The principal motive back of such work was the demand set up by the teacher; the manner of instruction was intrinsically uninteresting as few devices were employed. This type of instruction is known as formal drill. Its chief characteristic is that facts are drilled upon in isolation.

Rational Drill

In contrast to this we have to-day what is known as rational drill. It is an attempt to swing to the other extreme. Teachers are urged to teach by projects and situations that are close at hand. Children are not to do anything the reason for which they do not understand and which they do not consider of value to them. Extreme as this theory is it has had certain beneficial results. To-day practically every good teacher of reading begins with a combination of words that make sense instead of with the alphabet; the up-to-date teacher of music begins with rote songs; the progressive teacher of drawing permits the child to begin with a free-hand picture, crude and imperfect though it may be; and the modern teacher of arithmetic begins with the problems that are related to the experiences and interests of children. This is a great forward step in that it more nearly harmonizes instruction with the rational order of the learning process. It has given interest and zest to the daily work of the school. School discipline has been ameliorated because of this more modern way of teaching.

We are urged not only to begin each subject with situations and problems that are concrete and interesting, but we are told that the facility children need in reading will be acquired by multiplying the material they read; in singing by having them sing more songs; in drawing by having them draw more pictures; in arithmetic by having a multitude of problems solved. In other words, skill in technique is to be acquired by increasing the number of situations in which it normally occurs. For example, there comes a time in the singing exercises when attention to a bit of technique becomes necessary. This is presented and discussed in such a way as to assist the children in understanding it and in interpreting the song. This bit of technique appears the next day in that same song and on succeeding days in other songs. Similarly by solving many concrete problems the student in arithmetic becomes more and more automatic in his response to the number combinations.

The critics of this theory maintain that it requires too much time to produce the results demanded by business, and in this criticism there may be some truth. It seems that each of these contending theories contains an element of truth. The happiest and most fruitful combination would probably be as follows: After the pupils have learned to recognize and to use a sufficiently wide range of technical phases of a subject, these might be deliberately taken out of their natural setting and drilled upon formally. Such formal drill would have a distinct advantage over the old type of formal drill, viz., the children would see the use of the things they were being drilled upon.

Illustration of Rational Drill

A concrete illustration of this reconstructed form of drill work may aid in clarifying the description. The sixth grade children in the training school of a western normal school were deficient in the fundamental operations. It was decided to subject them to a daily drill in mental arithmetic. The teacher who was able to employ efficiently the greatest variety of devices was placed in charge of the work. The class recited in a room adjoining the one in which they studied. The recitation in mental arithmetic began the instant the children were assembled and seated in the recitation room. The teacher did not say, "Now, attention, children; we are going to have a little drill in mental arithmetic to-day. Is every one ready? Now listen carefully while I state the first example.'' On the contrary, the children understood that the recitation would start the instant the door closed. They knew the teacher would not waste several precious minutes in needless preliminaries. To be ready for her opening statement they were leaning forward on the outer edges of their chairs, ready to start just as a runner is at the firing of a pistol. The recitation moved aggressively. The teacher gave such examples as ''take 2, multiply it by 2, square the product, multiply by 4, take one-half that, 50 per cent of that, the square root of that.'' There were no pauses in her statement; she uttered these statements with the speed one uses in ordinary conversation. The very moment she finished the children were expected to be ready with the answer.

After a few weeks those who were interested in the experiment checked up on the children to note the number of mistakes they made minute by minute. It was noted that few mistakes were made the first two minutes, they were more noticeable during the third minute, very noticeable the fourth, equalled the number of correct answers the fifth, and exceeded the number of correct answers the sixth minute. On the basis of this evidence it was decided that only three minutes a day would be devoted to mental arithmetic, but it was to be three minutes of sixty seconds each, one hundred and eighty seconds in all. The experimenters knew full well that this amount of time did not agree with the amount set aside for such work in many courses of study.

The authors are convinced that the time that can be profitably devoted daily to oral arithmetic may be extended beyond three minutes. The extension of time depends upon the maturity of the pupils and the resourcefulness of the teacher. A great variety of devices is necessary if children are to be kept at their maximum speed and attention longer than three minutes.

Results of This Drill

This three-minute recitation became one of the "show things" of the school. The children looked forward to it with keen pleasure. In order that they might excel in it they learned the multiplication tables without any solicitation from any one up to 17, 18, 19, and 20. They could tell the cube of 16 as quickly as the average teacher can tell the cube of two. The writers have seen groups of two, three, and four of them in the corridor of the building, in vacant rooms, in the shade of trees on the campus, drilling each other. Perhaps some "progressive" may criticize the teacher for permitting them to learn the multiplication tables beyond twelve. They were so interested in the work that it would have been difficult to prevent them. The chief reason for stopping at ten or twelve is that it is conventional.

Whenever work of this type is advocated there is always some person ready to sneer at it—a person who remembers of some isolated case of a child thus drilled who never got into the high school, or who failed in all of his other duties, or who grew weaker in the reasoning problems in arithmetic. We want to assure the young teacher that a three-minute daily drill will not be accompanied by any such untoward results.

After this experiment had been carried on for some time an attempt was made to discover the effect of the drill on the written problems of the text. According to the testimony of the critic teacher more than sixty per cent of the written problems of the text had become mere oral problems to the children. This is a true measure of the value of such exercises. Moreover, the facility thus acquired was not lost during vacation; it carried over into each of the two succeeding years.

Although this experiment lacks some of the scentific value of others that we shall describe later, still it is sufficiently sound in all particulars to show the proper correlation of rational and formal drill, and also to show what remarkable results may be achieved. By devoting three minutes a day to rapid mental drill a teacher or county could achieve such distinction that their schools would be known throughout the land. It hardly needs to be said that personal distinction is not the end of such work. Three minutes a day will bring not only personal distinction; it will put the pupils in secure possession of the tools needed for higher mathematical work; it will so equip them as to break down one of the criticisms of the outside world.

Some Results Attained by Brief Drill

In this connection we wish to relate the results of another successful teacher who devoted three minutes daily to drill work in arithmetic. Sometime near the close of the year, three examples were given to the class as a test. They were:

1. Add 4587
8654
4879
6875
9897
8546
8465
7699
7967
4567

2. $7654219 \times 897 = ?$
$854796 \times 2078 = ?$
$27864523 \times 9376 = ?$

3. $89765342 \div 97 = ?$
$4275897245 \div 789 = ?$
$987007648 \div 654 = ?$

1. (Cont.) 7698
8765
7698
8765
7654
6574
5678
9876
5596
8945
7894
4894
4955
7644
8989
4554
6589

The three examples were given to pupils whose average age was 10¾ years. The following time was required in solving each example:

Problem	Time fastest pupil	First 25 pupils finished	Average time
1 .... 45 sec.	105 sec.	72 sec.
2 .... 110 sec.	270 sec.	184 sec.
3 .... 140 sec.	260 sec.	197 sec.

We are not attempting to justify the presence of such unusual numbers as appear in these examples. We recite the facts merely to show what may be accomplished by drill work without detrimental results to the other subjects of the curriculum.

Variations of Rational Drill

The two types of instruction we have been describing in this chapter have many variations when applied to other phases of arithmetic. One of the most common contrasts is revealed by the following incident:

A group of grade teachers was granted permission to visit a system of schools. Upon their return, one of them described a visit to a recitation in arithmetic. She said the teacher in charge had the day before assigned four

problems for the class to solve. When the recitation was
called, the teacher spent about five minutes in assisting
those children who had not been able to solve all the problems to discover their difficulties. Then she gave the
remainder of the time to the solution of other problems.
The critic, commenting upon this plan, said, ''I think she
should have had a child solve the first problem upon the
board; another child, the second problem; a third, the
third; and a fourth, the fourth. After they were all
placed upon the board, she should have had each of them
explained.''

Here we have two types clearly differentiated. How
shall we determine which is the better? Perhaps we should
ask, What is the test of efficiency in teaching material of
this kind? The test of efficiency in arithmetic is the solution of problems. If the children are able to solve intelligently all the problems in a proper assignment, it is *prima
facie* evidence that the teaching has been well done; if they
solve none of them, it is first-hand evidence that the teaching has been poorly done. One seldom finds either of these
extremes. What he does find is that some of the children
solve all of the problems, some none of them, and a few
some of them. If by a few well-directed questions the
teacher is able to make those who have totally or partially
failed conscious of the character of their difficulty, she
can then devote the remaining time to other matters. In
any event, she discovers those who are in need of individual
attention.

Much time is wasted in having children go over and over
problems that they have already demonstrated their ability
to solve. The ability of children to solve problems is a
direct measure of how well the teaching has been done.
If the time could be saved that is wasted in solving problems previously solved, many more problems might be

solved illustrating the application of the same principle; more time could be devoted to aggressive and effective drill work; and more attention could be given to the teaching of the methods of work involved in those new phases of the subject which are to come up later.

In this connection the writer is reminded of a recitation in a third grade in which the teacher imagined she was doing drill work in arithmetic. She had placed upon the blackboard a number of simple examples in addition. The class was called to "Attention"; one pupil was called by name; he ran lightly to the board; the teacher read the example while he copied it; he performed the addition as quickly as possible; the teacher smilingly nodded her approval; the pupil erased the work and ran back to his seat. This performance was repeated until each pupil had solved an example. The recitation began on time and closed on time, full twenty minutes having been consumed.

An examination of this shows that it had many of the marks of a good recitation. Each child had the opportunity to exercise himself physically—he went to and from the board; he exercised his mental activity, for he solved an example; he received the teacher's nod and smile of approval for the work done; perfect order and decorum prevailed throughout.

The weakness of the recitation was brought out in the conversation that followed at the rest period. The teacher asked the supervisor, who witnessed the recitation, what he thought of it. This supervisor, believing that his chief business was to improve instruction, asked the teacher why she did not send one pupil to the place where she had all of the examples copied, and then, as soon as the class had solved the first example and some pupil had stated the answer, have the boy at the board write it in its proper place. The supervisor wanted to know how long it would

have taken to have required all the pupils to solve all of
the examples. When the teacher said, "Oh, not more than
seven or eight minutes," the supervisor said, "Well?" A
moment later he said, "Well?" more interrogatively than
before. Then the teacher naïvely said, "But what would I
have done with the rest of the time?"

These illustrations are not overdrawn nor far-fetched.
They are descriptive of a type of work that can easily be
found in many places today. These traditional methods
are responsible for much waste in teaching. Such antiquated forms of instruction have led to conspicuously
wrong notions in regard to drill work. Successful drill
work means sharp, quick questioning and immediate responses; it should be accomplished in a short period of
time. If we eliminate the non-essentials from the text-
books and the uneconomical methods from the instruction,
we shall have an abundance of time to devote to things of
a more profitable character.

Accuracy and Speed

The two primary purposes of drill work are increased
accuracy and increased speed. Accuracy should involve
the securing of the correct result in the shortest possible
time with the minimum expenditure of energy. Inaccuracy results from attempting to do too much, from work
that is too difficult, from hasty and slipshod methods,
from a false attitude on the part of teachers. Teachers
sometimes encourage inaccuracy by praising work that is
wrong.

Laws of Habit Formation

Accuracy and speed seem to be closely related when the
work is done with the maximum degree of concentration.
One should work as accurately as he can, and as rapidly

as he can. By doing this he improves simultaneously in speed and accuracy.

Slowness in computation is due to any one of a number of factors. The most common cause is the lack of adequate training in counting and in manipulating the fundamental operations. When a child in the fifth or sixth grade finds it necessary to think what 3×7 are, he is either backward or has been subject to poor instruction. A response to such combination should come instantly. The lower nerve centers should provide the correct response.

The speed of the children is sometimes retarded because they are required to work too long upon supposedly concrete problems. This was one of the weaknesses of the Speer method. The writers knew a school in which practically no problems were solved during the first four years of the pupil's school life which he could not demonstrate with blocks or pictures. This sometimes resulted in keeping children so long upon the plane of concrete thinking that they lost or failed to acquire the power to handle abstract operations. Any good device of this kind when used to extreme becomes an evil.

A third factor which retards speed in computation is the failure to make the operations perfectly simple before drill is begun. At the outset the things to be drilled upon should be seen and understood in their normal situations. Then attention should be focused upon the operation itself. Mere repetition may result in a habit, but mere repetition is most uneconomical.

If one wishes to produce number habits with the least expenditure of energy and time, he should permit no exceptions,—the correct answer must be given every time. One never gets right results by praising wrong answers. Every response, every reaction, whether right or wrong, tends to impress itself indelibly upon the nervous system.

Another fact to be remembered is that mere exhortation on the part of the teacher does not produce habits in pupils. It is literally true that practice makes perfect, provided the practice is always upon the correct form. In the final analysis, one acquires particular skill only by practicing the things which give that skill.

The last important factor to be mentioned is that the periods between drills should be gradually lengthened, and not neglected. If we drill next week on what we have acquired this, then next month, then four months from now, then next term, then next year, and so on, there is little danger of children leaving school without acquiring many desirable number habits. It is for this very reason that texts should provide more liberally for reviews. Reviews that call for the mastery and fixing, as well as the wider application of facts and principles, are essential to scholastic attainment in every field, not alone in arithmetic.

Scientific Studies of Drill

One of the earliest studies of habit formation in arithmetic is that by Professor E. L. Thorndike, on "Practice in the Case of Addition" (*American Journal of Psychology*, 21; 483–486). The test was made with nineteen university students, eight men and eleven women. These added daily for seven days forty-eight columns of ten numbers each. No column contained any 0's or 1's. The time of each addition was kept in seconds, and a record was kept for the number of correct and incorrect results. The experiment showed four things: (1) that improvement in speed and accuracy was about equal; (2) the fact that adults can improve in a skill of this kind is good evidence that improvement in any intellectual trait is mainly the result of special training; (3) that practice improvement is greatest when one works up to his limit in competition with his own past record (this is the right incentive for special drills in regular school work); (4) that variability between individuals decreases with drill.

An extensive study of the effect of drill in the fundamental operations of arithmetic was made by Mr. Brown, one of the co-authors of this book.

How the Investigation was Conducted

Tests were given in the sixth grades of three different public school systems and in the sixth grade of a large private school. The total number of cases recorded in this study was 222; of these, 110 were boys and 112 were girls.

The three public schools examined are in the Central West. City C has a population of seven thousand; City M, of twelve thousand; and City D, of thirty thousand. The private school is in New York City.

The effects of the drill in fundamentals were shown by a comparison of sections subjected to the drill with sections of equal size and approximately equal ability not subjected to the drill but otherwise undergoing the same arithmetical instruction. The object was to determine the improvement made by the drill class upon its previous record and the improvement made by the non-drill class upon its previous record.

In a given class the tests were conducted at the same hour of the school day, in order to eliminate the time factor as far as possible.

Immediately after the first test was given in each school, half of the classes examined in each city were given five minutes' drill each day upon the fundamental operations in arithmetic,—addition, subtraction, multiplication, and division. The first five minutes of the recitation period in arithmetic were devoted to the drill work. The drill was partly oral and partly written, and the time was about evenly distributed among the four operations.

No special instructions were given to the teachers in charge of the drill sections, except that they were to emphasize both speed and accuracy in the four operations, and were to cover the same daily assignments in the text-books as the class that had no drill. The teachers of the non-drill classes were asked to give no formal drill upon any of the four fundamental operations during the time that this investigation was in progress. These instructions were carefully observed by the teachers.

The drill classes in each city were able to cover the same subject-matter of the text as the non-drill classes of that city. No special tests were given to determine the comparative excellence of the text-book work, but in every case the teacher in charge of a drill class reported that five minutes devoted to drill at the beginning of each recitation seemed to act as a mental tonic. It seemed to energize the pupils and to make them keen for the text-book work that was to follow. All teachers of drill classes reported an improvement in text-book work.

Formal drill work on the four fundamental operations had not been given prior to this investigation in any of the sixth grades examined. Whatever marked changes occurred in all of the drill sections that did not occur in the non-drill sections may reasonably be attributed to the results of the special drill.

Results of the Drill

If the number of problems worked in each test may be taken as a measure of the speed of the pupils, the drill class increased its speed by 16.9 per cent and the non-drill class by 6.4 per cent.

Since practically all of the pupils finished at least the first six problems in each test, a comparison of the records made on these six problems will give a basis for determining the relative accuracy. Measured by this standard, the drill class made a gain of 11.7 per cent in accuracy, whereas the non-drill class actually lost in accuracy (–1.8 per cent).

The largest gain made by the drill class was in division, 34.2 per cent, which was more than twice the gain made in division by the non-drill class, 15.4 per cent.

If we compare the gain made by the drill class upon its own record with the gain made by the non-drill class upon its own record, we find that the following results were attained:

Drill class gained 2.64 times as much as non-drill class on problems worked.

Drill class gained 2.72 times as much as non-drill class in addition.

Drill class gained 2.68 times as much as non-drill class in subtraction.

Drill class gained 2.21 times as much as non-drill class in multiplication.

Drill class gained 3.13 times as much as non-drill class in division.

Drill class gained 2.57 times as much as non-drill class in total number of points.

The drill classes made from two and one-fifth to three and one-tenth times as much improvement as the non-drill classes. It is worthy of note that the average age in the drill classes was exactly the same as the average age in the non-drill classes, being twelve and two-tenths years in each case.

In the following table the first test was given before the drill was begun, the second test was given immediately after the thirty days' drill, and the third test was given on the first day of the fall term, after a vacation of twelve weeks:

A COMPARISON OF THE RESULTS OF THE THIRD TEST WITH THE FIRST AND SECOND.

("I" indicates combined drill sections. "II" the non-drill sections.)

I did 26.4 per cent better than on first test and 4.1 per cent better than on second test in number of problems worked.

II did 9.8 per cent better than on first test and same as on second test in number of problems worked.

I did 25.4 per cent better than on first test and 6 per cent poorer than on second test in addition.

II did 7.7 per cent better than on first test and 3.7 per cent poorer than on second test in addition.

I did 46.2 per cent better than on first test and 6.7 per cent better than on second test in subtraction.

II did 20.4 per cent better than on first test and 6.4 per cent better than on second test in subtraction.

I did 31.3 per cent better than on first test and 1.5 per cent better than on second test in multiplication.

II did 11.1 per cent better than on first test and 2.2 per cent poorer than on second test in multiplication.

I did 36.7 per cent better than on first test and 7.3 per cent better than on second test in division.

II did 11.1 per cent better than on first test and 2.8 per cent poorer than on second test in division.

I did 31.7 per cent better than on first test and 0.2 per cent poorer than on second test in total points.

II did 12.16 per cent better than on first test and 2.29 per cent poorer than on second test in total points.

I did 5.2 per cent better than on first test and 0.6 per cent poorer than on second test in first six problems.

II did 3.7 per cent poorer than on first test and 1.3 per cent poorer than on second test in first six problems.

The results of the third test indicated that the superiority of the drill class was maintained over the vacation period. The "period of hibernation" served to increase the speed, while those who had not had the advantage of the drill worked no faster than on the second test. The non-drill section either made no improvement or did worse than on the second test in everything except subtraction.

The conclusions reached by Mr. Brown[1] were corroborated in all essential particulars by Dr. T. J. Kirby in a study entitled ''Practice in the Case of School Children,'' published by Teachers' College Bureau of Publications, 1913.

Mr. Kirby also found that a brief drill period produces better results than a longer period.

No investigation has yet been made to determine the relative efficiency of drill periods from one to ten or fifteen minutes, or whether the same length of period is best for each of the fundamental operations.

[1] A detailed account of Mr. Brown's investigation is given in the *Journal of Educational Psychology*, November and December, 1912.

CHAPTER IX

WASTE IN ARITHMETIC

This is an age of great commercial and industrial activity. The invention and improvement of numerous labor-saving devices, the improved facilities for transportation and communication, and the significant discoveries in numerous fields of scientific research have revolutionized commercial, industrial, and economic conditions in the United States within the last few decades. Efficiency has become the watchword, and efforts are being made to eliminate all possible waste of time, energy, and material. The efficiency engineer is a product of our modern conditions.

The progressive educator of to-day is interested in any investigation seeking to establish standards of efficiency that may be applied to the schools. Within recent years several school systems have been carefully examined and their efficiency has been judged by experts who had no immediate or personal interest in them. In so far as the equipment, organization, and instruction of the schools can be standardized and the efficiency measured, such investigations, when conducted with a genuine desire to learn existing conditions, are of real value.

It is a comparatively easy matter to establish standards of excellence for many material objects, and to classify more or less accurately the quality of a given product when compared with a standard product of the same kind. It is vastly more difficult, if not impossible, to set up exact standards of efficiency for all mental products. Some phases of school organization, equipment, and management can be measured with considerable accuracy when compared with recognized standards. Some of the products of the school cannot as yet be measured as accurately as we wish, but this should not deter us from applying standards of greater or less refinement to those that can be measured. The progressive educator will look with favor upon all attempts that are made to refine the standards of measurement now in use, and will welcome the establishment of other appropriate standards for the measurement of school activities and products. The teacher should constantly seek to establish a better balance between effort and result, a closer adjustment of means to ends. As far as is possible, standards for the evaluation of school activities should be used. Any investigation of subject-matter or of method in a given field that results in a discovery of ways and means of securing a better balance between effort and result is justified.

This chapter is devoted to a consideration of the extent to which waste can be eliminated in the teaching of arithmetic. The topic will be considered from the point of view of both the subject-matter and the methods of instruction.

Before a manufacturer can determine the percentage of waste involved in making a given product, he must have in mind the exact product that is to be made. If he wishes to determine the percentage of waste material, he must investigate the extent to which the raw material balances the output. There must be a consideration of means and of ends. A better adjustment of means to ends always means an economic gain. If we wish to investigate the sources of waste in arithmetic we must first establish more or less clearly the aims in teaching the subject. Unless the goals which are striven for are known, the degree to which current practice enables us to reach the goals cannot be determined. The degree of closeness of approach to the ideals to be reached is a factor of great importance.

We would justly condemn the manufacturer who had no clear idea of what he was attempting to produce. In the industrial world an efficient manager knows what he is attempting to do, and he sees the necessity for each step in the process. Each workman strives to accomplish a definite end. The part of the work that he does may be but a small part of that necessary to make the finished product, but it is the duty of each one to. do his part of the work to the best of his ability, whether it be the turning of a spool or the adjusting of a watch. It is not easy to determine what the final output should be in education, or what is the best method of securing the desired results.

Aims of Education

The aims of education are numerous, and the methods of attaining the aims must vary with the child. Some prominent writers on education enumerate five or six aims.[1] A variety of individuals makes necessary a variety of aims. Aimless teaching is wasteful teaching. Every teacher should have some working statement of aim, but this may vary.

Educators are not agreed in regard to all of the ideals to be striven for in the teaching of arithmetic, and it is not probable that complete agreement will be reached upon this point. Teachers of history, geography, grammar, and the industrial arts are not in agreement in regard to the ideals to be attained in their respective subjects. Numerous reasons for the teaching of arithmetic have been advanced, but most people will agree that in the main the subject has a two-fold justification. It is taught because of its practical

[1] See Bagley, "Educative Process," pp. 40-65; O'Shea, "Education as Adjustment," ch. 4 and 5.

bread-and-butter value, and because of the opportunity that it affords for training the pupil in concise, logical thinking. Arithmetic is by no means the only subject that serves these two purposes, but that it does serve them in a distinctive way, few will deny. No one questions the necessity of a mastery of certain fundamental number relations; all admit that a child must acquire a certain mastery of quantitative relationships. It is not an easy matter to evaluate the mental training that a pupil gets from arithmetic that he does not get equally from other subjects. The pupil is probably more conscious in arithmetic than in any other subject in the grades that he has come into contact with certain truth. When the mathematician Laisant was asked the relative importance of the utility and the culture value of arithmetic, he replied that it is like asking which is the more important, sleeping or eating,—the loss of either is fatal. The teacher who recognizes but one of these aims is not teaching most effectively. ''The practical side must concede to the disciplinary side by having its processes understood when they are presented, even though the child is not called upon to remember the reasoning. The disciplinary side must concede to the practical by selecting its topics in such a way as to give no false notions of business, and as to encourage the pupils to take an active interest in the quantitative side of the world about them.''[1]

Specific Aim of Arithmetic in Each Grade

The teacher should recognize not only the general purposes for which arithmetic is taught, but he should have clearly in mind the specific aim of the work in each of the grades in which he teaches. It is not wise to attempt to draw a fine distinction between the aims of the work in

[1] Smith, ''The Teaching of Arithmetic,'' p. 21.

each of the grades. These aims will be influenced more or less by local and individual factors, but in general certain purposes may be considered as of prime importance in the respective grades.

The object of the work of the first and second grades is to aid the pupils to image clearly the objects and groups of objects in proper number relations; to make clear and definite quantitative imagery. This may be accomplished through emphasis upon counting, upon addition and subtraction, upon simple estimates and comparisons; games, rhymes, drawing, construction work, and the like.

It is the specific object of the third and fourth school years to make the pupil proficient in addition, subtraction, multiplication, and division with whole numbers and certain fractions, as a basis for the work that is to follow. Emphasis should be placed upon accuracy and speed in both oral and written work. By the use of simple problems adapted to the experience of the pupil, there should be developed the ability to interpret simple number relations and to reason from simple data. The pupil's appreciation of quantitative relationships in the life about him should be considerably developed in these grades, through continued use of measurement and estimates, and a study of the simple tables of compound numbers.

In the fifth and sixth grades the mechanics of the fundamental processes with integers and fractions should be thoroughly mastered. There should also be increased emphasis upon the solution of problems. Considerable time should be devoted to the selection of appropriate processes for solving problems. Every effort should be made to develop independent thought by strengthening the judgment in the selection of correct processes. The use of appropriate checks should become habitual in these grades, and this will do much to increase confidence and to bring pride in accomplishment that is so desirable. Problems in these and all other grades should be, as far as possible, within the pupil's experience. The data of geography, history, science, and manual training may be utilized to a considerable extent.

The pupil who passes the age of eleven or twelve without the ability to perform the fundamental operations with facility and a relatively high degree of accuracy is quite likely to be handicapped in these respects throughout life. Teachers of the fifth and sixth grades have a great responsibility upon them, and the pupil's success in arithmetic in the following grades depends to no small degree upon the results that are secured in these grades. Many a child is sent from these grades a cripple in his work in arithmetic. Whenever possible, individual attention to specific needs should be given, and much loss of time in later grades may thus be avoided.

It is the purpose of the arithmetic of the seventh and eighth grades to give a mastery of percentage and its modern applications, of mensuration, ratio and proportion, and of square root; to afford a thorough and comprehensive review of the entire subject of arithmetic; and to give the pupil some knowledge of the elements of algebra and geometrical construction. In these grades especially, the larger aspects of social, industrial, and economic life should be emphasized. The solution of numerous problems and the application of appropriate checks should be continued, and every effort should be made to develop independence of judgment and confidence in results.

Elimination of Topics

The arithmetic of the recent past included several topics which should have been eliminated because they no longer served any practical end. Traditions of the school have always exercised a powerful influence in retaining topics in arithmetic after their period of usefulness has passed. Certain topics, like certain organs in the human body, tend to persist after their original functions have been outgrown.

In former years, the defense offered for the retention of any topic in arithmetic that was no longer of practical value was that it possessed a disciplinary value, and this was thought to justify it. To-day we recognize both the practical and the disciplinary value of arithmetic, but we do not believe that any topic should be retained merely because of its disciplinary value. We seek the maximum of mental discipline in the topics that have some practical value. If a practical topic is properly presented, the disciplinary value will be obtained from it. Both insight and skill may be acquired from the study of such a topic.

One of the great sources of waste in arithmetic to-day is due to the fact that we have not yet eliminated all topics that find no application in present-day practice. The established traditions of the schools, the tendency of teachers to teach as they were taught, and the fascination of some of the old types of problems, tend to perpetuate the obsolete. It is wasteful to teach any topic in the grades that the pupil will not use in some way, either outside of the classroom or in the school work that comes later. Even in the latter case a topic should not usually be presented until some use is to be made of it. To say that the mind of the pupil is developed by the study of an obsolete topic does not justify its retention, because the same or a greater development may be secured by the study of topics of genuine practical value. Any course in arithmetic that does not look forward to the life of the twentieth century rather than back to the nineteenth is wasteful. Any course that does not seek to apply arithmetic to the problems of daily life is not fulfilling its full purpose.

What Topics Should Be Eliminated

The demands of the present insist that the topics enumerated below be omitted from a course in arithmetic in the grades:

1. Greatest common divisor and least common multiple of all numbers not readily factored, and all work involving the Euclidean method. The terms of fractions to-day are relatively small and easily factored, or the fractions are expressed decimally. Prior to the beginning of the seventeenth century, business practice demanded a knowledge of the long-division form of the greatest common divisor. It is not improbable that the subject of greatest common divisor and least common multiple will be omitted entirely from courses in arithmetic in a few years. The common fractions taught should be those actually used in business and in practical life.

2. All obsolete tables in denominate numbers and all tables that are of use to the specialist only. Troy and Apothecaries weight are not of importance to most people.

3. All problems dealing with compound numbers of more than two or three denominations and all reduction from Troy to Apothecaries weight should be omitted.

4. All work in circulating decimals should be omitted; the topic should be studied as a part of infinite series in algebra.

5. All applications of percentage that do not conform to present-day practices should be omitted. True discount should not be taught. The numerous state rules on partial payment should receive no consideration unless they are in general use in the community. The types of negotiable paper should be those in common use among business men.

Equation of payments is not of practical value to-day to anyone except a few specialists in foreign commercial transactions.   Annual interest should not receive much emphasis.

6. Cube root should not be taught in the grades.

7. Progressions have no practical value in arithmetic. The theory of progressions is distinctly algebraic.

8. Compound proportion has been largely replaced by unitary analysis. Even simple proportion is of less importance than formerly, because of increased emphasis upon the simple equation and analysis.

9. Problems which require long and involved solutions or answers should be omitted.   Those listed below, copied from an old arithmetic, are extreme illustrations of this.

Typical Problems from Musgrove and Wright's "British American Commercial Arithmetic," published in 1866, Toronto, Canada.

1. Simplify

$$\frac{\frac{1}{2}+\frac{1}{4}+\frac{1}{8}+\frac{1}{12}+\frac{1}{16}+\frac{1}{32}+\frac{1}{64}+\frac{1}{128}}{\frac{1}{2}+\frac{3}{4}+\frac{7}{8}+\frac{15}{16}+\frac{31}{32}+\frac{63}{64}+\frac{127}{128}}$$

2. Reduce the common fraction $\frac{1}{49}$ to a decimal.

   Ans.   .020408163265306122448979591836734693877551

3. A, in a scuffle, seized on $\frac{2}{3}$ of a parcel of sugar plums, B caught $\frac{3}{8}$ of it out of his hands, and C laid hold on $\frac{3}{10}$ more; D ran off with all A had left, except $\frac{1}{7}$, which E afterwards secured slyly for himself; then A and C jointly set upon B, who, in the conflict, let fall $\frac{1}{2}$ he had, which were equally picked up by D and E.   B then kicked down C's hat, and to work they all went anew for what it contained; of which A got $\frac{1}{4}$, B $\frac{1}{3}$, D $\frac{2}{7}$, and C and E equal shares of what was left of that stock.   D then struck $\frac{3}{4}$ of what A and B last acquired, out of their hands; they, with some difficulty, recovered $\frac{5}{8}$ in equal shares again, but the other three carried off $\frac{1}{8}$ apiece of the same.   Upon this,

they called a truce, and agreed that the ⅓ of the whole left by A at first should be divided equally among them; how many plums after this distribution had each of the competitors?

Ans.   A had 2863; B, 6335; C, 10294; and E, 4950.

Enriching the Course

Much waste has been eliminated from the course in arithmetic by the omission of obsolete topics, but great care must be exercised to insure that some equally useless topics are not substituted for those that have been eliminated We are attempting to-day to enrich our modified curriculum by introducing numerous problems from practical life. We are seeking to secure a better mastery of the fundamental operations through greater emphasis on systematic drill and through greater emphasis on oral arithmetic.

Unnecessary Computation

There is waste of time and of effort in arithmetic whenever a process is carried further than the data upon which it is based would justify. No result in mathematics can be more accurate than the data upon which it is based. Many pupils never appreciate the absurdity of finding the interest on a given sum to within a millionth part of a cent, or of finding the circumference of a carriage wheel to within a millionth of an inch. For most practical computations a result computed to two or three decimal places is sufficiently accurate.

Time is wasted because of the performance of unnecessary operations. It is wise *never to multiply until you are forced to, and never divide until you are obliged to.* The following problem will illustrate this point: Required to find the radius of a circle equal to the combined area of two circles of radii 6 and 8 inches, respectively.

Necessary Work:

The area of the required circle $= 36\pi + 64\pi = 100\pi$

The radius of any circle $= \sqrt{\dfrac{a}{\pi}}$

The radius of the required circle $= \sqrt{\dfrac{100\pi}{\pi}} = 10.$

In solving such a problem many pupils would multiply both the 36 and 64 by the numerical value of π, then add the products. This multiplication is unnecessary.

Cancellation frequently enables one to save time in the solution of a problem. For example:

1. If 29 bushels of potatoes sell for $20.88, what will 31 bushels sell for at the same rate?

SOLUTION:

$$\frac{31 \times \$20.88}{29} = \$22.32$$

Time Devoted to Arithmetic

For many years arithmetic was the foremost study in the curriculum in determining the pupil's standing. It is still an important subject in this respect, but its position of preëminence has been taken by English or Reading. Pupils formerly devoted about one-third of their school time to the study of arithmetic. As early as 1850, other subjects which were demanding admission to the curriculum sought to curtail the time given to arithmetic. About 1875 arithmetic was displaced as the dominant factor in determining promotion. To-day fifty of the leading cities of the United States devote an average of 15.2 per cent of the school time to the subject. The decline in the time devoted to arithmetic is not due to a feeling that arithmetic is less worthy, but to the demands of other subjects for admission. Arithmetic will probably maintain its position of prominence for many years, but it will eventually surrender a portion of its time to such other topics as are sufficiently thought out and systematized as to justly claim a share. A sufficient number of school hours are still devoted to the study of arithmetic. What the schools need is not more time for the subject, but a clearer realization of the purposes for which arithmetic should be taught and a better adjustment of means to ends. We need to economize the time that is now devoted to the subject rather than waste it by aimless and unsystematic instruction.

Applications of Arithmetic

It is wasteful to teach arithmetic without sufficiently emphasizing the applications of the subject to life. The study should help the pupil to interpret life from the quantitative point of view, and unless it does this the subject is not fulfilling its full purpose. A pupil should be taught his arithmetic so that he can apply it to problems outside of the book. Too many pupils can solve complicated problems about prisms, pyramids, and cones, but cannot find the volume of their father's coal bin or the amount of dirt that must be removed in digging a cellar. Unless the facts of the classroom are interpreted in terms of facts outside the classroom the teaching of the subject is not efficient. There should be a direct and immediate relation between many of the problems of the arithmetic class and outside experiences that demand similar knowledge. Many of the experiences of the pupil in the school should vitalize the experiences outside of the school, and many of the experiences outside of the school should be utilized to clarify the work of the school. Pupils are too often required to solve problems of whose use and application they can have no clear conception.

Unity of Arithmetic

It is wasteful to teach arithmetic as a multitude of unrelated topics when in reality the subject contains only a few distinct processes. The pupil should appreciate the fact that the same process may appear under various phases of the subject. One who has no appreciation of the unity and the simplicity of arithmetic has failed to grasp the full significance of the subject; arithmetic to him is a maze of rules and processes. The teacher should show the pupil how, from a few definitions and fundamental processes, the entire science is developed step by step. To present a new topic without showing its relation to those that have preceded is wasteful teaching.

The Thought Side of Arithmetic

There is waste in the teaching of arithmetic because not enough attention is given to the development of the pupil's power to reason. Not enough training is given in the selecting of the proper process to be used in a given situation.

Mechanical processes are of great importance in arithmetic, and our pupils have not become too expert in them; but in many schools mechanical ability has been developed largely at the expense of the ability to see relations and to think, to select appropriate processes with certainty. Frequently too small a proportion of the arithmetic period is given to developing the thought side of the subject. The pupil should be able to perform the fundamental operations with integers and fractions with reasonable facility and with a high degree of accuracy; but it is necessary also that he should know when the performance of a given process is necessary in the solution of a given problem. If the judgment of the pupil is to be developed in arithmetic, the teacher should exercise great care that he does not hamper it by the imposition of numerous rules. Much of mechanical teaching is wasteful teaching.

Assignments

There is much waste of time and of effort in the study of arithmetic because the teacher does not make the assignment with definiteness and precision, and as a result of this the study hour is one of blind grouping and discouraging failure, resulting in disgust for the study and in dawdling habits. All assignments are intended to afford growth toward independence and initiative in thought and action. The immediate purpose of an assignment varies, but in general its purpose is to guide the pupil in preparing for whatever the new recitation will present, to direct him toward the accomplishment of what he can do by himself, and to save time in the recitation proper.

The teacher should always have a definite and satisfactory reason for assigning each lesson. This reason should be based on the teacher's knowledge of the pupils, their interest and needs, and on a knowledge of the lesson. A well-made assignment saves time in succeeding recitations and makes the recitation more profitable than it could possibly be without the preparation of the child's mind that is afforded through the assignment.

In the lower grades especially, all assignments should be made with great care and precision. As the pupils grow in power they may sometimes be permitted to suggest the assignment, which will be discussed and amended by the teachers. Some classes, late in the grades, may occasionally be trusted to make the assignment. This affords an excellent means of strengthening the selective power and of encouraging originality.

A good assignment will put the pupil in a mood to work on his lesson, and will stimulate him to attack and solve the problems before him.

Explanations

It is questionable how much explanation should be given by the teacher when the assignment is made. Some assert that the teacher should not point out the difficulties, but should leave the child free and unhampered by any assistance. This view will probably entail a great waste of time and energy in most classes. It often leads to the magnifying of trivialities, and results in unorganized knowledge.

It is probably good practice in the lower grades for the teacher to explain nearly all the difficulties, and in the upper grades to make the assignment in such a way as to assist in removing them. One measure of the value of an assignment is the degree of interest aroused and maintained in the class in its study apart from the teacher. If the teacher knows both his pupils and the subject to be taught, he should be able to arouse the interest of the pupils in the subject.

If in the recitation, or even when making the assignment, the teacher works up to an absorbing point and then leaves it in suspense, the pupils will return to the subject with keen interest during the study hour.

Just enough time should be devoted to the assignment to make it clear and definite—to remove the obscurities and certain difficulties, and to create enough interest to insure further study. Sometimes this can be done in a few minutes; at other times it will require most of the recitation period. The assignment is a place for raising problems and creating interest in those problems. Haste and slovenliness are to be avoided, because they are destructive to energetic effort on the part of the pupils. No general directions can be given as to the proper time to make an advance assignment. Usually it should be made at the beginning of the recitation period. If the assigning of the advance lesson is postponed until the close of the period, teachers frequently find that they have not allowed themselves sufficient time, and the result is a hurried assignment, lacking both definiteness and precision.

A teacher has no right to expect a pupil to prepare with precision an assignment that was not made with care and with definiteness. It sometimes happens that the advance assignment is to depend largely upon points to be developed during the recitation period or upon the extent to which the recitation has been satisfactory. Under such conditions the teacher cannot know the appropriate length for the assignment or what features should be made most prominent until late in the period. Some excellent teachers always make the advance assignment at the close of the period, but the custom is not one that can be generally practiced with the best results. Individual assignments, such as telling a pupil that he will be held responsible for the solution and explanation of a given problem at the next recitation, may properly be made in the midst of the recitation.

Methods of Study

One of the greatest sources of waste in education is found in improper methods of study. Many earnest pupils who are anxious to study do not know how to do so "independently, intelligently, or economically." The teacher should do all that he can to assist the pupil to form good habits of study. Two books which discuss the general problem of how to study should be noted, Sandwick's "How to Study and What to Study" and McMurray's "How to Study." The most extensive investigation of the various methods of teaching pupils how to study mathematics is to be found in Part I of the Thirteenth Year Book of the National Society for the Study of Education. These books, especially, contain numerous suggestions on the general problem, and some of them apply to the study of arithmetic.

Pupils should learn to overcome difficulties without the aid of the teacher. A spirit of self-reliance and of tenacity of purpose are excellent assets for any pupil. Interest in one's work may be initiated and augmented by consistent application to a task that may originally have been without interest. Many a pupil has become tremendously interested in his mathematics by persevering over tasks that were at first uninteresting.

If a pupil is to prepare his mathematics assignment with economy of time and of effort, the conditions under which he studies should be such as to admit of a high degree of concentration upon the work in hand. The habit of concentrating the attention is of great value in studying any subject, and especially mathematics. Many pupils do not seem to realize the necessity of selecting, when this is possible, an environment that is conducive to study, and that will permit concentration upon the task. If parents and pupils can be convinced that under proper conditions twice the work may be accomplished in half the time, a great gain has been made.

Waste in Mechanical Procedure

There is waste of the recitation period because the teacher permits pupils to form dawdling habits of work, especially at the blackboard. This can usually be avoided by attention on the part of the teacher. In many classes much time is wasted in such procedures as the calling of the class roll, passing to the blackboard, and the distribution of papers. The writers visited a class in arithmetic in which the teacher consumed eight of the forty minutes allotted for the recitation in the calling of the roll and the distribution of papers. During this time no comments or suggestions were made in regard to any of the work. This teacher wasted about 20 per cent of the recitation period in doing things that should have been done with less confusion in one-tenth of the time. The recitation that followed reflected the dawdling habits and unsystematic procedures of the teacher. Subsequent observations revealed the fact that the teacher wasted at least 20 per cent of the recitation period each day in such ways as have been indicated. As the teacher is, the class is. A teacher who does not work in class with economy of time and of effort will soon find that many of the pupils have developed habits of work quite similar to his own in that subject. If a teacher wishes the maximum amount of work to be accomplished during the recitation period, even the minute details of the work must be planned with this end in view. Not every pupil in a class will adjust himself to the ideals of the teacher, but it is certain that unless the teacher himself has proper standards of work the general average of the class in this respect will be materially lowered.

Reviews and Examinations

Some educators would eliminate reviews and examinations from school work. They say that such eliminations will delight the American boy and that no compulsion will be necessary to induce him to attend school. School life will be a continuous round of pleasure, and there will be no work to make Jack a dull boy. Such radical proposals are without social or psychological warrant. Sociology informs us that it is the part of wisdom to hold onto those things that long and successful experience have proved to be worth while. Psychology informs us that drills and reviews are necessary in the fixing of habits and in organizing material.

It is a fundamental principle of psychology that if one is equally hazy about two things formerly learned, one months ago and the other recently, it requires a smaller amount of effort to recall and fix the older of the two than it does to recall and fix the one more recently acquired.

Reviews properly distributed pay tremendous dividends in the economy of mental life. Surely the old practice, which provided for the automatic mastery of the fundamentals, should not be discontinued without a hearing. Many things still need to be reduced to the automatic or to be firmly fixed in the mind by means of drills and reviews. No true scholastic attainment or efficient education is possible without a reasonable use of reviews and drills.

Our pupils to-day are in more danger of suffering from intellectual starvation than from mental dyspepsia. A school is organized and maintained in order that it may be instructed, and reviews are an essential part of efficient instruction. The social or educational reformer who would eliminate all drill and all reviews is striking at one of the essentials of instruction.

In individual lessons attention is focused primarily upon the various parts of the subject-matter rather than upon the unity of the parts. A good review transfers the attention to the larger relations. The unity of a subject is likely to be lost in the multitude of detail, and a comprehensive and thorough review is necessary, not only to fix the important points more firmly in the mind, but to produce that thoroughly organized and well-articulated knowledge which is such a valuable asset. Reviews should appeal to both the eye and the ear.

The ingenuity and skill of the teacher are frequently taxed to the utmost in order to devise ways and means of instilling a high degree of interest into a review lesson. It is not necessary that the subject-matter and the method of presentation be the same as when the topics were previously studied. Indeed, the advance that has been made will frequently suggest a broader view and a better method.

Thorough and comprehensive reviews are especially necessary at the close of the year's work. Investigations on this point indicate that such reviews are economical in that they save time in the work of the succeeding year. A brief review of the principal features of the previous year's work should also be given during the first week or two of the school year.

It is the custom to-day in some localities to decry examinations and to urge that they be eliminated from the school. It is doubtless true that examinations in the past have frequently been given with little or no thought as to their real purposes.

Examinations should usually be comprehensive in character: their purpose should be to aid the pupils in organizing and correlating the knowledge acquired into a coherent and systematic unit. The examination should be regarded as a means to an end, and not as an end in itself. Too often the examination is regarded simply as a means of testing the memory for specific and unrelated facts. It is proper that a part of an examination should be devoted to this purpose, but the important facts to be tested should be the ability to organize and to apply.

Examinations given by the teacher, if well planned and

wisely constructed, frequently aid the teacher to discover individual or class weaknesses, and to furnish a basis for further instruction. Such examinations, if given with due regard for the welfare of the pupils, are justifiable.

Examinations given by superintendents, principals, or supervisors usually have as their aim the setting of standards of work throughout the system, the testing of the efficiency of the course of study, and the discovery of individual weakness. Such examinations, if properly conducted, may be made an effective administrative device. Broad-minded teachers usually welcome the opportunity to have their work tested.

The present tendency is to decrease the frequency of examinations and to use more care in their preparation. The marks received on examinations formerly determined whether a pupil was to be promoted. This is not true in many schools to-day. The practice of giving "finals" is now uncommon. The memorites type of examination is being replaced by those which require old knowledge to be utilized not in isolation but in new situations. More questions demanding thought, judgment, and choice are given to-day than in former years.

PART THREE

CHAPTER X

PRIMARY ARITHMETIC

Preliminary Statement

The arithmetic which children are first put to work upon should be closely related to their lives. Arithmetic is one of the agencies consciously prescribed by society for giving children control over a particular phase of their environment. This control, however, cannot be economically or advantageously acquired unless children are provided with normal situations that provoke natural reactions. A genuine mastery of number in the early school years is gained by using it in a concrete manner—in the construction of play-houses, in games, in weighing, measuring, and counting those objects and relationships that represent the daily enterprises of child life.

Much of the teaching of arithmetic has not been dominated by this ideal. In the past, instruction was characterized by the memorizing of rules, formulas, and examples. Little attempt was made to relate the materials and methods of arithmetic to social and industrial life. The advent of a multitude of new subjects, such as nature study, gardening, manual training, domestic economy, and agriculture, had much to do in converting the old reflective school into a more active school. The presence of studies in the curriculum is no longer justified by their hypothetical mind-training value. Now they are justified by the number of relations of identity they have with the world.

This influence has spread across and modified instruction in the older subjects. The new movement is seen in arithmetic in the elimination of materials no longer socially serviceable, in the addition of materials closely related to the business practice of current life, and in the rationalizing of instruction through object work.

Nature of Primary Arithmetic

Primary arithmetic has been restricted by common consent to a knowledge and mastery of the fundamental operations as expressed in integers and in fractions. It is not vocational in the sense that its processes are named after particular occupations. This division of the subject-matter is due largely to the preparatory character and value of these fundamental processes. They are the intellectual tools which all must use in their later life, no matter what occupation they choose. Reform tendencies in method have reconstructed the traditional presentation of these tools by rules and formulas to the more rational discovery of their nature in concrete problems.

Dominance of Methods in the Teaching of Arithmetic

Mathematical instruction was long dominated by the logical theories and forms of organization of men of science. Each fact was taught to show its relation to other facts of the subject, and not to connect it with some common operations of life. Courses of study were constructed in the office of the superintendent and handed over ready-made to the classroom teacher. Those special and personal modifications so essential in reconstructing the experiences of children were seldom encouraged. For this reason teachers taught as they had been taught by men of science, or as they had been told to teach, and not as the exigencies of the situation demanded. Instruction was logical, not psychological. The measure of the value of a fact was its relation to the other facts in an organized scheme, and not its recurrence in daily life.

Logical vs. Psychological Method

The logical method or arrangement of material represents the adult or the scientific point of view, while the psychological method or arrangement represents the manner in which children normally approach and master the situations of a subject. When materials are logically arranged there is always some central, organizing integrating principle. When they are psychologically arranged there is a distinct attempt to present materials to harmonize with the psychological conditions of childhood. The spiral or concentric circle plan is built upon the theory that instruction should proceed from the simple to the less simple, and from the less simple to the more complex. The facts and skills first presented, whether they are addition, subtraction, multiplication, or division recur later in more complicated form, and reappear again and again as the circles widen. Materials are thus adjusted to suit the various psychological stages or maturity levels of children.

These methods have had a reactionary influence upon each other. Naturally the older one is the logical. It is so ingrained in our practice that teaching has suffered from its baneful effects. The spiral plan, like every reform movement, tended to swing practice to an unnatural extreme. Compromise was inevitable. Now some phases of arithmetic, for example, parts of denominate numbers are taught topically, while the fundamentals, whether integers or fractions, are taught spirally.

Formal Discipline

This scientific justification for mathematical instruction, which came down from the university and took root in the primary schools, was supplanted by the doctrine of formal discipline, which invited support and justified both the logical form of organization and the presence of old materials in the curriculum on the ground that they trained the mind and hence were of value in later life, no matter how remotely they were related to the processes of daily life. This doctrine afforded a sanction for the retention of obsolete materials and antiquated methods. Although more people were interested in building and loan associations and in insurance than in the greatest common divisor, cube root, or partnership, the teacher argued vigorously for the retention of the latter on the ground of their mind training value.

The disciplinary defense gradually began to break down before the demands of a commercially prosperous public. A commercial age is always a period of great educational transition and advance. It seeks to justify the practices of the school in terms of needs and conditions outside of school. Business men exalted instruction in those particular arithmetical processes that were needed in business practice, and discouraged others. Business utility operated both as an eliminating and as a selective agency. Impractical and obsolete materials were discarded and new materials were introduced. The restricted point of view of the business man is now being transformed by a wider social point of view which calls for that instruction that conserves our common human obligations rather than trains for a specific vocation.

Time to Introduce Arithmetic

Opinion differs as to the proper time for beginning the study of arithmetic. In some schools a definite part of the school day is set apart during the first school year for the study of numbers. In other schools such work is postponed until the second school year, and in a few schools no formal study of numbers is taken up until the third school year. In the more advanced European countries arithmetic is studied during the first school year and the tendeney seems to be in that direction in this country. Until about one hundred years ago arithmetic was never taught to pupils just entering school. Many think that any regular and systematic attention to number is premature in the first school year and that the time should be spent in widening the pupil's activities and knowledge through elementary nature study, garden making, games, drawing, and constructive exercises. They contend that it is legitimate to introduce only so much of number work as will serve a definite end in these subjects. The work in arithmetic, they maintain, should be incidental and not formal and systematic in the first school year. Whenever a quantitative relationship presents itself in any school activity they would attempt to make it clear. They maintain that such a plan tends to develop number ideas naturally and does not force them and hence makes the subject more attractive to the child.

Pupils are asked to note the number in the class, the number absent or tardy; to compute the number of pencils, or books needed for the class or for a given row. Games offer numerous opportunities for introducing number ideas because of the necessity of keeping score, etc.

Others contend that when a pupil comes to school he is just as anxious to learn the fundamental relation of numbers as to learn to read, and that it is as unwise to postpone one as the other. Too often incidental teaching becomes accidental and perfunctory teaching, and a pupil finishes his first school year with his number ideas but little advanced. It is wise to motive the number work in the first and in all other grades, but many teachers wrongly assume that interest in a subject for its own sake is a motive that should not be considered. Most young children naturally like arithmetic, but many of them soon come to dislike it simply because of poor teaching in the lower grades. One of the great wastes in arithmetic lies just here. The interest of many pupils in the study of numbers is deadened and as interest wanes accomplishment fails and the pupil finds that he has developed a positive dislike for the subject. In general, the methods of teaching in the primary grades are more carefully thought out and systematized than in the intermediate grades, but a great responsibility rests upon the teachers of the primary grades for giving a pupil as good a start as is possible for him in the subject.

Correlations

Innumerable opportunities arise in other subjects to teach number. The objects used should be those in which the children are interested. Much comparatively dull work has been done in the past by using objects having had only an adventitious interest. Highly decorated cards, curiously carved animals and queer looking sticks do not serve the purpose best. The objects used should be changed frequently. It is stupid to teach all the number operations by using beans.

Children must learn to find their seats, their place in the line, to get six erasers, three pieces of crayon, to distribute five pencils or four cards, and to turn to a given page. Opportunities of this kind for the teaching of counting will not be overlooked by the resourceful teacher.

Correlation With Constructive Work

Every lesson in constructive work affords many opportunities for teaching numbers. Every lesson upon the triangle, square or rectangle is to some extent a lesson in numbers. When such materials are used the teacher should make no attempt to differentiate the four fundamental operations.

A multitude of interesting constructive devices have been invented for teaching number, each of which is of value not only as an occupational device, but because of the intellectual by-products resulting from its mastery. Among other occupational devices listed for the first grade are: square seed box, seed envelopes, table, the three chairs of the three bears, the three beds, basket, sled, soldier cap, boat, folding basket, square prism, cube, triangular prism, pyramid, closed seed box, trunk, cradle, bath tub, candy box, match stand.[1] Most of the above exercises can be taught without using the ruler, but they cannot be taught without training the number sense of the children. As soon as the pupils have acquired some skill in handling the ruler the number of objects they can make is increased. Some of those described in Mr. Worst's book are: paper chairs, paste trays, postage stamp holder, cornucopia, mat, woven basket, thread winder, color exercises, pin holder.

The Use of Objects

Successful primary work to-day is almost intuitively associated with object teaching. No good primary teacher would think of attempting to teach numbers without a supply of objects. If these are judiciously selected they serve as the school's best substitute for those outside situations that attract the native tendencies of children. At any rate they are the school's one best means of making concrete the abstract numerical concepts that up to the middle of

[1] The steps used in making these objects are described in a book entitled "Constructive Work," by Edward F. Worst, Superintendent of Schools, Joliet, Illinois, published by A. W. Mumford, Chicago, Illinois.

the nineteenth century were taught by purely memoriter methods. Pestalozzi has been called the father of object teaching. This is correct only in the sense that he restored the use of the object to the schools. As children gain in power of independent thought, that is, in ability to organize and condense their experiences, the number of objects used in teaching decreases; but they never entirely disappear, for they are always of value in comprehending and interpreting new situations.

Dr. Suzzallo's discussion of the materials of objective teaching is particularly illuminating. His first indictment is the artificiality of the materials employed. "Primary children count, add, etc., with things they will never be concerned with in life. Lentils, sticks, tablets and the like are the stock objective stuff of the schools, and to a considerable degree this will always be the case. Cheap and convenient material suitable for individual manipulation is not plentiful. But instances where better and more normal materials have been used are frequent enough in the best schools, to warrant the belief that more could be done in this direction in the average classroom. The 'playing at store,' and the use of actual applications of the tables of weights and measures are cases that might be cited."

His second indictment is that the materials have too restricted a range. The range of materials is almost directly dependent upon the resourcefulness of the teaching staff. When teachers and supervisors know what they want and need in teaching, there will be comparatively little difficulty in inducing boards of education to buy it.

His third indictment is the limited use of the materials on hand. "It is too frequently the case that the teacher will treat the fundamental addition combinations with one

NOTE—For an elaborate discussion of "The Teaching of Primary Arithmetic," see *Teachers' College Record*, March, 1911, Suzzallo.

set of objects, e. g., lentils. In all the child's objective experience within that field there are two persistent associations—'lentils' and the relations of addition. A wide variation in the objective material used would make teaching more effective, particularly with young children.''

His fourth indictment is that the materials of objective teaching have been too narrowly interpreted. He advocates a more extensive use of pictures, both in text-books and in teaching, and of geometric figures and diagrams which are of value, though in a more restricted way, in extending the range of concrete experiences.

His final indictment is that unsympathetic and conservative teachers use objects in a highly artificial and disorganized way. On the other hand the resourceful and progressive teacher secures some unity even in the materials she handles; she uses a game, some community activity, some childish interest, permits the children to play ''store,'' or conductor on a street car, etc.

The wasteful effects of undue emphasis upon unrelated objects may be shown by a description of an actual recitation. The number fact to be taught was $6+3$. The children were led to see that

$$6 \text{ blocks} + 3 \text{ blocks} = 9 \text{ blocks}.$$

This was arrived at by putting 6 blocks in one group and 3 blocks in another, combining these groups and then counting, after which the pupils in turn repeated the ''story.'' The process was then repeated with marbles and the children told the ''story,''

$$6 \text{ marbles} + 3 \text{ marbles} = 9 \text{ marbles}.$$

Then followed beans, toothpicks, pennies, shoe-pegs, peas, apples, crayons, pencils, rulers. Then pictures of cherries, oranges, nails, oblongs, square, and chairs were used. Following this they passed to remember objects and repeated the appropriate "story" for adding birds, trees, boys, lemons, horses, potatoes, and bricks.

In this particular school all of the primary lessons were taught in this way throughout the entire year. It took a year—*a whole year*—to teach the number facts and relations from 1 to 10. Had these children been ignorant of these particular facts a few illustrations would have supplied them with an adequate basis for making the proper generalizations.

Use of Games

The principal device currently used by teachers in fixing the fundamental arithmetical concepts, is the group play of children. Games permit of group work where the attention is predominatingly spontaneous. Consequently it is not uncommon to find children mastering the number operations under the free and wholesome atmosphere of plays. The fascination of the game, which is due partly to the element of chance, reduces the symptoms of fatigue and supplies a motive and zest for the work of the following day. The otherwise artificial recitation of the examination sort is thus transformed into a source of pleasure. One of the practical benefits early realized is that children soon become skillful enough to play such games independently of the teacher.

List of Games

1. Dominoes
2. Parchesi
3. Crokinole
4. Bean bag
5. Card games
6. Flinch
7. Hop scotch
8. Ten pins

9. Tag   12. Lotto
10. Odd and Even   13. Jackstones
11. Marbles

Around the Circle.   The digits are arranged in the form
of a clock face upon the blackboard.   Any number of these
digits may be used.   A digit is then placed in the center
and the numbers are multiplied by it as rapidly as possible.   If a pointer can be made to swing on a pivot, the
game may be varied by taking the numbers on which the
pointer rests.   The game can also be varied so as to relate
to addition and subtraction.

Bird Catcher.   A variation of this well-known game is
to have children sit in a circle, each taking a number. The
pupil in the center gives easy examples.   When the result
is the number of any pupil in the circle that pupil holds
up his hands.   When a number already agreed upon is the
result, all hold up their hands.

Buzz.   The members of the class count in turn.   When
a given number or any of its multiples is reached, they
say "buzz" instead of number.   Those pupils fall out who
miss or who hesitate in their answers.

Climb the Ladder.   Draw a ladder on the blackboard.
Put a combination on each step, letting each pupil climb
until he falls.

Hide and Seek.   Combinations are placed on the blackboard with the different parts missing—as, $14 + 3 = (?)$;
$14 + (?) = 17$; $(?) + 3 = 17$; $4 \times (?) = 24$.   In getting the answer the pupils unconsciously repeat the combination and
form the desired habit.   This is a very desirable form of
play, and it is characteristic of the kind of game that a
teacher can make for herself.   Instead of the symbol $(?)$
it is often advantageous to use X, thus giving a valuable
algebraic form that will be helpful at a later time.

Morra. This is a very old Roman game and is played with great enthusiasm by Italians, old and young. A group of children stand or sit in a circle. Each extends, at a given word, all or any of his fingers. An immediate estimate is made as to the total number. All are added then to see who is nearest right.

Roll the Hoop. Draw a wheel on the board, or use an actual wheel. Place number combinations for spokes, and, by stating results in order, play that you make the wheel "turn" as fast as you can.

The use of games opens the way for children to handle the objects. Instead of being passive listeners with the teacher manipulating the objects as she demonstrates a solution, they thus become active participants in the process.

Formal and Rational Methods

Objects and games in the primary grades were at first but the remote evidence of the quickening effect of critically scientific methods, which had filtered down from the higher institutions through the grades in diluted form. Primary teachers were not keen in grasping the new idea; they did not see that a wider use of well selected materials would assist them in correcting the dogmatic methods of instruction then in vogue. This was no new attitude on the part of the teachers; as a class they have never solicited new subjects and have seldom warmly welcomed new methods. These things have come in under the influence of outside pressures that could not be withstood. In this particular instance the lower schools were held in the grip of memoriter methods, facts were learned with but the mini-

For an instructive and suggestive list of number games and rhymes see Teachers' College Record, November, 1912, published by Columbia University.

mum attempt at rationalization. The employment of induction in the more strictly scientific fields spread across and produced as its counterpart "developmental" instruction in the grades. As the currents of imitation spread from the higher to the lower schools, primary teachers, not yet conscious of the modes of inductive thinking, seized upon its most obvious and material features, and began to use objects for purposes of illustration. It was not until later that teachers sought to have children rediscover or invent through inductive thinking or developmental instruction the valuable things the race has discovered or invented. Under the influence of this idea it was not uncommon for teachers by a skillful use of the question and answer device to do practically all the thinking for the class. Another and far more unfortunate result was the practice of multiplying objective illustrations of a fact long after the children understood the fact. The attempt to rediscover every relation and every feature about a number fact by object teaching, was as deadening as the more purely memoriter methods which made but an indifferent attempt at relating facts to the experiences of children.

Emphasis in instruction has changed from the mass to the group, from the class to the individual, from the transmitting of vicarious experience to face to face experience, from a single appeal to a variety of appeals. The educational center of gravity has shifted from subject matter to child. The teacher thinks less of her function as a transmitter of knowledge by repetitious exercises and more of the necessity of breathing personality into dead materials, of multiplying the points of contact between the materials and the minds of the children, of permitting children to learn through their own experiences rather than through the dogmatic statement of book or teacher. The teacher tries to stimulate a feeling of need in the minds of the children. True leadership as expressed in educative methods rather than coercive methods characterizes the spirit of modern instruction. Drill as shown in another chapter is rational rather than formal.

Counting

Attention will now be directed to certain specific phases of arithmetic. The most fundamentally important thing for beginners is to learn to count. Counting consists in thinking objects serially related and, later, in thinking the symbols that represent those relationships. Consequently, it is not synonymous with measurement, for that consists of comparing or measuring objects alike in nature—as blocks, two lines, two sides of a room, some shorter, some longer. A quantitative idea of the world of nature can be gotten only by comparing and measuring, but the most elemental ideas of number are gotten by counting. The most fundamental mathematical concept depends upon a recognition of the one to one correspondence between objects. The expression of such correspondences is counting. Counting is as old as the race. Primitive people use a variety of objects in counting—pebbles, sticks, stones, toes, hands, fingers, shells, grains of corn, notches on sticks, knots in strings. Their idea of number was expressed through the correspondence of the thing with the number of notches (cut in a stick) or knots (tied in a string), and names to express the correspondence were limited or wanting. With the invention of names counting began.

The facts of number are fixed through counting. Children learn that 17 is less than 18 and more than 16. In other words, the place relations of numbers are fixed by counting; particularly if the pupils learn to count forward and backward. The operations of addition, subtraction, multiplication, and division are attempted refinements of complex forms of counting.

Number Symbols

The next problem after children have acquired the number concepts through the use of objects is that of teaching them to read and write number symbols. The problem that the teacher has at this point is precisely the one she has in teaching beginning reading. The pupils already recognize by ear the number names, and are reasonably facile in communicating them by speech, but they do not know the figures by sight and cannot write them. Sight recognition must precede, logically and chronologically, the writing of the symbols. At the beginning there must be a definite association of the symbol and the written form with the idea through objects. This association has usually been made by the children themselves, so that formal instruction of this character may not be needed at all. Those courses of study that set the limits from 0 to 10 for the first term or year, and from 10 to 20 the second term, and 20 to 100 the third term, are not only mechanical and illogical but they destroy interest, because they do not take advantage of the mental equipment of the children. It is contrary to the tenets of psychology to expect one to have the same interest in a skill that has already become automatic. The process of reading and writing figures is interesting in itself. The prevailing tendency, however, is to associate the number series with a game or some interesting activity. If any instruction in the place relations of numbers apart from objective associations is necessary, it should follow the objective presentation of the numbers. Instantaneous recognition and response should follow the display and perception of figures.

Notation

The Arabic notation is a method of combining numbers, giving them value according to the place they occupy, letting each figure indicate ten times as much as the one to the right. It is a simple device, an artificial method of giving to a few figures a remarkable power. The reading of numbers of large denominations presupposes a knowledge of notation. When pupils have learned the number language up to ten, this may be extended rapidly from ten and one to ten and nine, and from two tens to nine tens and nine, and so on. When they can count and write to 100, the teacher may safely introduce units, tens, and hundreds; when they can read and write to 1,000, she may introduce thousands, and so on. A detailed application is seen in addition. When children can add single columns up to nine or less, and can write to one hundred, they may be taught the adding of double columns without carrying. In teaching them the reason for writing one figure under another, the teacher may proceed by saying, "Let us see how many tens there are in 48." The class discovers that there are four tens and eight units; this explains why 48 is written as it is. The teacher next says, "I want to write 41 under 48. Let us see how many tens there are in it." The children count ten, twenty, thirty, forty—four tens and one unit. It is then easy to see that if we want to add all of the ones (units) together and all the tens together, the most convenient arrangement would be to write them in separate columns.

Roman Numerals

Very little attention should be given the Roman numerals, for two reasons: (1) They are rapidly going out of use, and (2) they are difficult to learn and more cumber-

some to write than Arabic numerals. Some years ago
Professor Robert M. Yerkes of Harvard conducted a number of experiments to determine the time required to write
Roman numerals and the number of errors made in using
them. He found that it takes three and one-third times
as long to write the Roman numerals from 1 to 100 as the
Arabic, and the chance of error is twenty-one times as
great; it takes three times as long to read the Roman
numerals from 1 to 100 as the Arabic, and the chances of
error are eight times as great. In view of their limited
use, there is no excuse for teaching any Roman numerals
except those from one to twenty, fifty, one hundred, five
hundred, and one thousand.

CHAPTER XI

THE TEACHING OF THE FUNDAMENTALS

The Five Operations

There are but five simple operations to be learned. "Each of them involves one or more quantities measured by some unit. If we have a simple quantity—for example, a 24-inch line—we can repeat it a given number of times, say, three times, the result being 72 inches; or we can divide it into a number of equal parts, say, three, the result being 8 inches. If we have two quantities measured by a common unit, we can combine them, thus finding how many of the common units there are in the whole; or we can find how many more of that unit are in the larger than in the smaller; or we can find how many of the smaller quantity there are in the larger. All of arithmetic consists in doing one or some combination of these operations, and it is essential that the child have so much experience in handling objects to perform them that he shall think their meaning and know better than to try to multiply 5 inches by 3 inches, or to add 5 and 4 inches."[1]

Now, the society for which our schools are preparing does not demand the mastery of a logically arranged series of mathematical principles. It does demand a thorough knowledge of a few things, and their applications. For example, it demands ability to count, to read, and to write numbers; also accuracy and rapidity in the fundamental operations. To this end every well-ordered course will

---

[1] Fiske Allen, "Is Arithmetic a Science of Numbers or of Symbols?" *The Kansas School Magazine*, Emporia, November, 1912.

provide an abundance of systematic drill in these processes, a drill that will be continued throughout the upper grades until the desired standards of accuracy and rapidity are secured.

The Serial Relation of Numbers

Some of the best courses of study demand entire familiarity with the number scale to 120 before any work in formal addition is begun. Care must be taken that the transition from things and images to symbols shall be gradual. The counting and recognition of the number scales involves knowing that any number and 1 more gives the next number in the scale, and that the number immediately preceding any number in the scale is one less than the number, and that one less than any number is the number immediately preceding it. The school soon affords numerous opportunities for teaching counting; for example, the number of children in the room, the group or class; the number absent; the chairs in the room, pencils in the box; the selection of certain objects from other objects; the counting of objects both seen and touched; the counting of sounds (taps); the repetition of an act. This may be followed by more abstract work when children count forward and backward, and later by tens, fives, twos, and fours. It is clear that such counting involves both addition and subtraction. When these have been begun in a more formal way, the number scale may be extended, and other numbers used for cumulative counting.[1]

The series idea which is the basis of the number concept may be abstracted very early in the school experience of the average child.

It is unnecessary to teach every number objectively. A

[1] See California State Course of Study.

pupil soon comprehends that 16 or 23 or any number occupies a certain place in a series. He does not think of it objectively. For this reason the number scale should be taught and memorized as such.

The structure of the number scale may be shown in the grouping involved in counting by tens.

0	10	20	30	40, etc., etc.	90	100
1	11	21	31	41, etc., etc.	91	
2	12	22	32	42, etc., etc.	92	
3	13	23	33	43, etc., etc.	93	
4	14	24	34	44, etc., etc.	94	
5	15	25	35	45, etc., etc.	95	
6	16	26	36	46, etc., etc.	96	
7	17	27	37	47, etc., etc.	97	
8	18	28	38	48, etc., etc.	98	
9	19	29	39	49, etc., etc.	99	

Counting by ones and tens may be extended to counting or adding by twos, three, fours, fives, etc. These exercises lie at the foundation of good work in addition. Oral drills of this kind should be given until all sums of this character can be given accurately and rapidly.

In the reading of whole numbers, do not allow pupils to use "and," as one hundred *and* eighteen. This should be read one hundred eighteen. The "and" will be needed later to indicate the decimal point.

Addition

Although addition grows out of counting, care must be taken to avoid counting in adding. Both the oral and written forms of the combinations must be memorized perfectly. Drill until the sums can be given at sight.

The order of presentation of the combinations should be:

(1) objects, (2) representation of the combinations—objects not present, (3) recalling the combinations—values memorized, (4) concrete problems without the use of objects. Readiness in giving the result of any two numbers should be considered of first importance. Attention must be called to the fact that addition is the process of combining units of the same denomination. Such quantities as dollars and gallons cannot be added; th an only be set down. Such quantities as yards, feet, ld inches can be changed to like units.

In the case of forty-five of the eighty-one addition combinations which require automatic mastery, the sum is not greater than ten. These forty-five combinations follow:

$$
\begin{array}{ccccccccc}
1 & 2 & 3 & 4 & 5 & 6 & 7 & 8 & 9 \\
1 & 1 & 1 & 1 & 1 & 1 & 1 & 1 & 1 \\
\hline
2 & 3 & 4 & 5 & 6 & 7 & 8 & 9 & 10 \cdot
\end{array}
$$

$$
\begin{array}{cccccccc}
1 & 2 & 3 & 4 & 5 & 6 & 7 & 8 \\
2 & 2 & 2 & 2 & 2 & 2 & 2 & 2 \\
\hline
3 & 4 & 5 & 6 & 7 & 8 & 9 & 10
\end{array}
$$

$$
\begin{array}{ccccccc}
1 & 2 & 3 & 4 & 5 & 6 & 7 \\
3 & 3 & 3 & 3 & 3 & 3 & 3 \\
\hline
4 & 5 & 6 & 7 & 8 & 9 & 10
\end{array}
$$

$$
\begin{array}{cccccc}
1 & 2 & 3 & 4 & 5 & 6 \\
4 & 4 & 4 & 4 & 4 & 4 \\
\hline
5 & 6 & 7 & 8 & 9 & 10
\end{array}
$$

$$
\begin{array}{ccccc}
1 & 2 & 3 & 4 & 5 \\
5 & 5 & 5 & 5 & 5 \\
\hline
6 & 7 & 8 & 9 & 10
\end{array}
$$

```
 1   2   3   4
 6   6   6   6
 —   —   —   —
 7   8   9  10
```

```
 1   2   3
 7   7   7
 —   —   —
 8   9  10
```

```
 1   2
 8   8
 —   —
 9  10
```

```
 1
 9
 —
10
```

A convenient way to express addition combinations is:

	1	2	3	4	5	6	7	8	9
1	2	3	4	5	6	7	8	9	10
2		4	5	6	7	8	9	10	11
3			6	7	8	9	10	11	12
4				8	9	10	11	12	13
5					10	11	12	13	14
6						12	13	14	15.
7							14	15	16
8								16	17
9									18

There are other ways in which the important combinations may be related. For example, there are the old-fashioned addition tables:

$$1+1=2$$
$$1+2=3$$
$$1+3=4$$
$$1+4=5 \text{ etc.}$$

The combinations may be arranged on the basis of the sums. For example:

$$1+1=2 \qquad \left.\begin{array}{l}1+6\\2+5\\3+4\end{array}\right\}=7$$

$$1+2=3$$

$$\left.\begin{array}{l}1+3\\2+2\end{array}\right\}=4 \qquad \left.\begin{array}{l}1+7\\2+6\\3+5\\4+4\end{array}\right\}=8$$

$$\left.\begin{array}{l}1+4\\2+3\end{array}\right\}=5$$

$$\left.\begin{array}{l}1+5\\2+4\\3+3\end{array}\right\}=6 \qquad\qquad \text{etc.}$$

Column Addition

The next two pages are taken verbatim from the course of study of the Los Angeles City Schools:

2	3	3
6	5	4
2	2	3
3	9	3
4	3	4
3	6	5
3	2	6
4	3	2
3	4	9
2	5	3

Use the combinations now mastered in building columns for adding. An examination of these illustrative columns will show how columns can be constructed and extended upward as far as the teacher likes.

In beginning column addition with children in the primary grades, place the following column on the board.

3	
4	
3	
9	32
3	23
6	20
2	14
3	12
4	9
3	5
2	

Take the chalk, and, beginning at the foot of the column, say: "Two, three,—five," pointing to the numbers as named, and write the 5 to the right of the 3. Then say, "Five, four,—nine." Write the 9 to the right of 4. Then say, "Nine, three,—twelve," and write the 12 to the right of the 3. Then continue, "Twelve, two,—fourteen," writing the 14 to the right of the 2, and so on until the column is added. At each step have the children, collectively or individually, repeat after you each statement. Drill the pupils until they can go through this without error. If there is any hesitancy about the combinations, point to the combination above, so that they may learn where to find the correct form if they should forget.

After this process and language form is established, write similar columns on the board for each pupil, with instructions for him to do the exercise himself. The teacher should pass from one to another, hearing each give the form. As a pupil finishes, let him exchange examples with another pupil, first erasing the side columns. To avoid confusion it is well to write two or three examples in excess of the number in the class, so that no pupil need wait. As a further convenience, it may be helpful for the pupil who finishes a column to write his name underneath it. The teacher, passing around, later erases the answer and the side columns, and writes "C" (correct) or "X" (wrong) after his name. The place is then ready for another pupil.

With a few pupils there will be a continual tendency to make mistakes in the left-hand figure, to write 42 instead of 32, etc. This means that insufficient work has been done on the number scale. Suppose, as in the illustration given, the pupil writes 42 instead of 32. To correct this, several methods are at the option of the teacher.

3
4
3
9   42
3   23
6   20
2   14
3   12
4   9
3   5
2

(1) She can go back for more drill in the decades, then make the application to the difficulty in hand. (2) She can have him write, in ascending column, the number beginning with 23, until the next 2 is reached. (3) She can draw a line under 23, and ask, "What 2 next above 23?" (*Answer*, "32.")

After the combinations already mentioned have been mastered, and every child can work out the side columns of any column of figures built up out of these combinations, readily and without mistake, the same combinations, in their reverse form, should be treated in like manner.

The purpose of this is to drill the pupils in learning new combinations and in visualizing the end figures of the successive partial sums. After this form has been mastered, the teacher should continue addition without writing the sums at the side, and train the pupil to add without this help. In starting this it is well to require the pupil to add directly, thus: five, nine, twelve, fourteen, twenty, twenty-three, etc. If he makes mistakes, have the pupil, in imagination, go through the form of the partial sums in the side column, without actually writing them. First attempts will be slow, but a few exercises will cause him to depend upon his own visual imagining. Proceed in the same way to add other columns in review.

In all this early work, the child should never be permitted to perform any work in addition at his seat, but always at the board, in full view of the teacher. Children, if allowed the time, will fall back into the habit of counting up the sums serially. It is a mistake to think that children will "outgrow" this habit, once it is formed. Changing one's habits is not so simple a matter as this. To prevent this habit from being formed, the teacher must first give in columns only those combinations which the children have first learned thoroughly, and, second, always insist that the work be performed at the board and in full view of the teacher. Do not permit the child to stop and "think." He either knows the sum or not. If he shows the least hesitancy he must either be told the answer or be permitted to look at the combination involved in its answer. For this purpose the combinations should, with their sums, always be written on the board in full view of the child.

Concert Recitations Versus Individual Tests

Concert work is good, but it should not be employed exclusively, for many children are thereby made dependent in their work. Again, if a teacher uses it too generously, she cannot know what the individuals are capable of doing. In addition work, the teacher must keep in mind the fact that her class will not proceed uniformly in the acquisition of the work, and that in consequence she must provide some way to give much individual instruction. The principal of each building should keep in close touch with each of his teachers in the work. To do this, he should take individuals from the classes into the office, or into some convenient room, and there test them as well as drill them to supplement the work of the teacher. He should know when a teacher has completed the study of a given group of combinations, and determine, through taking the class, whether it is ready to proceed to the next group.

Each successive group should be treated in the same way, except for the side columns, which may be discontinued. The work, however, would better be conducted at the board, for reasons already given. The teacher should prepare columns of figures within the limit of the particular group which she is treating, then dictate these to the class at the board. Each child writes the column, beginning at the bottom and going toward the top, and of course adding in the same direction, in order that the combinations may be as intended. After all the groups have been studied and columns are given, with the combinations arranged heterogeneously, it does not matter whether the child adds up or down. In fact, it is a good check on the work to have him add both ways. In this dictation work, after the children have obtained a sum, several should be called on to add aloud; then several called on to add the same columns, but with 2 tens, 3 tens, 5 tens, 6 tens, etc., prefixed to the lowest number. This gives drill in the upper reaches of the number scale without the additional work of rewriting the numbers. Columns should continually be given which incorporate and use the combinations of groups already learned. The pupil's advance into new ground should be very slow, in order that he may master the old very thoroughly. The chief merit of this, as of any other system used in teaching combinations, rests in its thoroughness. The child must pass by easy and carefully graded steps from the simple to the difficult. At every step of the way, the teacher must keep well within the powers of the child. Men succeed in this, not so much by reason of past failures as because of past successes. We like to do the things we can do well. Just so with the child; he gets a pleasurable emotional reaction from doing things at which he is successful. This is the chief value, as well as pleasure, of review work—it perfects technique, and becomes pleasurable in proportion to the child's success in the doing of it.

Motive for Speed

At first the emphasis must fall on accuracy and neatness. To attempt to secure too much speed at first, leads to inaccuracy. But gradually, as the work becomes more reflex, "speeding-up" exercises should be given. Here is a place for the right kind of emulation, such as is found in contests among classes, or among individuals. There is a tendency among the advocates of "soft pedagogy" to disparage rivalry in the schoolroom. History shows us, however, that this motive has been a powerful factor in every line of social and individual progress. Because rivalry has a selfish, anti-social side, it does not follow that it lacks a noble and helpful one. It is not well to foster emulation to the extent done by the Jesuits, who went so far as to pair off all the boys of a school, making the individuals of each pair rivals in everything pertaining to school work. It can be used safely, however, in pitting class against class, or, if tactfully done, individual against his fellows of the same class. Within these limits emulation will prove itself a powerful schoolroom incentive.

Habituate the Carrying Processes

Early in the work with the groups, numbers of three or four figures each can be dictated if the teacher desires it. However, to do this without introducing unfamiliar combinations, the teacher must think out the numbers, taking into consideration the figure to be "carried." Efficiency in adding demands that the processes become reflex. The adult mind, when adding columns of figures, or when subtracting one number from another, is absolutely devoid of even the feeling of the concrete. To begin in the first grade, tying splints into tens, and these tens into hundreds, is interesting, perhaps, but it gives no working ability, and we question its value in giving so-called insight into number. As a matter of fact, the mechanical process of "bringing down the 2 and carrying the 1," to the child, is just as much an objective thing as would be a bundle of splints, and, besides, it happens to be right along the line of the child's future as well as present need. One does not need to know anything about the mechanism of an adding machine to operate it successfully, nor of a watch to read the dial plate.

Multiple Column Addition

Single column addition is naturally and logically followed by two, three, and four column addition, and as soon as numbers can be written and read to ten thousand and above, addition may be rapidly extended to any number of columns. Double column addition should be carried on at first without carrying. Short columns like

20	22	26	22	12	24
30	30	30	32	21	32
				33	13

should be used at first. These may be followed by longer columns and by columns that involve "carrying." A good device for teaching "carrying" and also for securing accurate work, recommended and employed by the Civil Service, is herewith illustrated:

$$
\begin{array}{r}
83245 \\
6278 \\
5312 \\
246 \\
7582 \\
\hline
23 \\
24 \\
14 \\
21 \\
8 \\
\hline
102663
\end{array}
$$

Subtraction

Addition and subtraction are so closely related that they may be taught simultaneously, particularly if the additive or Austrian method in subtraction is employed. It advocates "subtracting by adding" rather than by "taking from" or by "borrowing." Subtracting by adding is the method used by the expert cashier, by all money changers, by the business world. A number of successful plans for teaching subtraction, such as the complementary method, the borrowing and repaying plan, simple borrowing, the left-to-right plan, and the "Austrian" method, are described by Smith.[1] The most recent of these is the "Austrian or the making change" method; it dates from the sixteenth century. It consists in finding what number must be added to the subtrahend to make the minuend. If 6 is to be subtracted from 13, one thinks what number must be added to 6 to make 13. This plan has two decided advantages: (1) it avoids the necessity of learning separate tables for addition and subtraction, and (2) it increases speed and accuracy in addition, as subtraction forms are immediately converted into addition associations. "The meaning, the applicability, and the visual form of addition and subtraction are still different. Only the process of remembering and using the fundamental operations is the same."

Ordinarily pupils should be required to master only one of these methods. A pupil who has already become habituated to a certain method of subtracting in one school should not be required to learn a new method when he changes schools.

The same idea is easily applied to subtractive processes

[1] *Teachers' College Record*, Columbia University, 1909.

involving "borrowing." In solving the following exercise, for example, the child would say,

$$\begin{array}{r} 8341 \\ 6456 \\ \hline 1885 \end{array}$$

"six, *5*,—eleven; six, *8*,—fourteen. Five, *8*,—thirteen; seven, *1*,—eight."

The criticism has been made that the addition idea can not be used in the subtraction of fractions and in the subtraction of dates, but this criticism is not based on fact.

The subtraction combinations are the same as the addition combinations; the lower figure in each instance being equal to or smaller than the upper one. A simple arrangement of them is:

1	2	3	4	5	6	7	8	9
1	1	1	1	1	1	1	1	1

	2	3	4	5	6	7	8	9
	2	2	2	2	2	2	2	2

		3	4	5	6	7	8	9
		3	3	3	3	3	3	3

			4	5	6	7	8	9
			4	4	4	4	4	4

				5	6	7	8	9
				5	5	5	5	5

					6	7	8	9
					6	6	6	6

$$
\begin{array}{ccc}
7 & 8 & 9 \\
7 & 7 & 7 \\
\hline
\end{array}
$$

$$
\begin{array}{cc}
8 & 9 \\
8 & 8 \\
\hline
\end{array}
$$

$$
\begin{array}{c}
9 \\
9 \\
\hline
\end{array}
$$

Another arrangement may be based upon the minuend; as,

$$
\begin{array}{ccccccccc}
9 & 9 & 9 & 9 & 9 & 9 & 9 & 9 & 9 \\
9 & 8 & 7 & 6 & 5 & 4 & 3 & 2 & 1 \\
\hline
\end{array}
$$

$$
\begin{array}{cccccccc}
8 & 8 & 8 & 8 & 8 & 8 & 8 & 8 \\
8 & 7 & 6 & 5 & 4 & 3 & 2 & 1 \\
\hline
\end{array}
$$

$$
\begin{array}{ccccccc}
7 & 7 & 7 & 7 & 7 & 7 & 7 \\
7 & 6 & 5 & 4 & 3 & 2 & 1 \\
\hline
\end{array}
$$

$$
\begin{array}{cccccc}
6 & 6 & 6 & 6 & 6 & 6 \\
6 & 5 & 4 & 3 & 2 & 1 \\
\hline
\end{array}
$$

$$
\begin{array}{ccccc}
5 & 5 & 5 & 5 & 5 \\
5 & 4 & 3 & 2 & 1 \\
\hline
\end{array}
$$

$$
\begin{array}{cccc}
4 & 4 & 4 & 4 \\
4 & 3 & 2 & 1 \\
\hline
\end{array}
$$

$$
\begin{array}{ccc}
3 & 3 & 3 \\
3 & 2 & 1 \\
\end{array}
$$

$$\begin{array}{cc} 2 & 2 \\ 2 & 1 \\ \hline & \\ & 1 \\ & 1 \\ \hline \end{array}$$

Every device indicated under addition may be and should be used in teaching subtraction. We list two additional ones for drill work. Supply figures omitted:

Addition

6	6	2	3	4	5	2	5	1
10	11	8	9	7	12	9	11	6

Using subtrahend, ask how many makes the minuend; as,

9	8	10	13	21	5
-4	-3	-7	-6	-7	-3

Multiplication

In multiplication in arithmetic we have a number to be taken as an addend a given number of times. Multiplication involves three numbers, the multiplicand (the number to be repeated); the multiplier (the number showing the number of repetitions); and the product, showing the result.

The multiplicand and the product must be of the same denomination. This is not violated by such statements as $\$3 \times 2 = \$6$, because this should be read $\$3$ multiplied by $2 = \$6$.

The multiplication tables developed slowly. It was not

the custom for them to be learned by the student until after the seventeenth century. Human ingenuity devised a number of interesting schemes to make it unnecessary to master the tables. The "Sluggard's Rule" illustrates how elementary products were obtained. If one wished to multiply 8 by 6, the following form was used:

$$8 \times 2$$
$$6 \quad 4$$

The 2 and 4 are the respective deficiencies of 8 and 6; i. e., $10 - 8 = 2$ and $10 - 6 = 4$. The multiplication when completed would appear—

8×2 The four of the result is $6 - 2$ or $8 - 4$, and the 8 of the result is 4×2

$\dfrac{6 \quad 4}{4 \quad 8}$ Thus $8 \times 6 = 48$. Such a scheme requires no multiplication beyond five.

Millions of peasants in Russia to-day secure the correct result in multiplication without knowing any of the tables. Any table may be arranged to show the relationship between addition and multiplication; for example, the table of twos:

									2
								2	2
							2	2	2
						2	2	2	2
					2	2	2	2	2
				2	2	2	2	2	2
			2	2	2	2	2	2	2
		2	2	2	2	2	2	2	2
	2	2	2	2	2	2	2	2	2
2	2	2	2	2	2	2	2	2	2
2	4	6	8	10	12	14	16	18	20

The following will be found a convenient arrangement in connection with the tables and in counting by twos, threes, fours, fives, etc.

It will be noted that the figures of the second column (Figure I) represent the counting by twos. This is also true of the second row. Those in the third row (or column) ascend by a constant addition of 3; those in the fourth by 4, etc. The diagonal from A to C contains the figures 1, 4, 9, 16, etc.

FIGURE I

A											B
1	2	3	4	5	6	7	8	9	10	11	12
2	4	6	8	10	12	14	16	18	20	22	24
3	6	9	12	15	18	21	24	27	30	33	36
4	8	12	16	20	24	28	32	36	40	44	48
5	10	15	20	25	30	35	40	45	50	55	60
6	12	18	24	30	36	42	48	54	60	66	72
7	14	21	28	35	42	49	56	63	70	77	84
8	16	24	32	40	48	56	64	72	80	88	96
9	18	27	36	45	54	63	72	81	90	99	108
10	20	30	40	50	60	70	80	90	100	110	120
11	22	33	44	55	66	77	88	99	110	121	132
12	24	36	48	60	72	84	96	108	120	132	144
D											C

FIGURE II

1	2	3	4	5	6	7	8	9	10	11	12
2											
3											
4											
5											
6											
7											
8											
9											
10											
11											
12											

The following number game is based upon the above arrangement.

Have several pupils at the board (or seats) each prepare a large square containing the small squares as indicated in Figure II.

The teacher says "8." The pupils write this either under the 4×2 or the 2×4. The teacher says "40." This may be placed either under 5×8, 8×5, 4×10, or the 10×4; the teacher continues to state numbers and the pupils put them in the proper squares. The pupil who first fills all of the squares in a row or a column or a diagonal wins the game. The speed with which the numbers are announced by the teacher should be adjusted to the class, but care should be taken to insure that the numbers are announced rapidly enough to keep the pupils working near the limit of their ability.

Several variations of this game will suggest themselves to most teachers.

Division

Division is the analytic process of finding one factor when the other is known. A distinction is not infrequently made between division by partition and division by comparison or measurement. In partition a unit is to be divided into equal parts, the number of which is known and the size of each is required. The following is a problem in partition. Seventy-two books of the same kind cost 288 dollars; how much was that apiece?

Numerical solution: $\frac{1}{72}$ of \$288 = \$4.

Oral analysis: Each book cost one seventy-second of \$288, or \$4.

In division by measurement we have a whole to be divided into equal parts, the size of which is known and their number required. A problem in division by measurement is: A mother divided 8 cakes among her children, giving each two, how many children has she? Numerical

solution: $2 \overline{\smash{)}\,8 \text{ cakes}}$ cakes $4 =$ the number of children.

Oral solution: She has as many children as 8 cakes are times 2 cakes, or 4.

The numerical expression for measurement or division may be developed from subtraction in the same way that multiplication was developed from addition. Give a problem involving the division of a number (say 18) into equal parts, the size of each part being (say 6). Show that this same thing may be expressed more briefly.

$$
\begin{array}{r}
18 \\
-6 \\
-6 \\
-6 \\
\hline
0
\end{array}
\qquad
\begin{array}{r}
3 \\
\hline
6)\overline{18}
\end{array}
$$

Short Division

Before beginning the more formal work in division a child should be able to tell instantly how many times 2 is contained in every number up to 20, and should state the remainder, if any. The same principle holds for 3 and 30, 4 and 40, etc. The most elementary facts of short division are taught in the learning of the tables. The problem grows a little more complicated when the dividend is composed of several digits. But when the pupil is ready for this type of division, he should be familiar with the system of notation in use. With this as a basis, the explanation of the division of 6234 by 2 is relatively simple. The quotient for purposes of illustration may be expressed as $3000 + 100 + 15 + 2$. This form of expressing the quotient, however, should not be encouraged or indulged in for any considerable length of time. On the other hand every pupil should be able to express any quotient in such terms whenever occasion demands it. Whatever constitutes current practice gives us the clue to the form to be used in teaching short division. In the problem $2)\overline{6234}$ there is only one difficulty, i. e., the division of 2 into 30 or into 3 tens and 4 units. This difficulty is more apparent than real, for this very division has already been taught in connection with the tables. The truth is there is no real difficulty in teaching the facts and processes of short division, if the tables have been properly learned. The difficulty, such as there is, is in notation.

```
  112
21)2352
  21
  ‾‾
  25
  21
  ‾‾
  42
  42
```

The procedure is properly related to knowledge already known if one begins with 11 into 121 and gradually enlarges the dividend. A list of examples, prepared by Mr. O. T. Corson, editor *Ohio Education,* illustrates our point:

$12221 \div 11 = ?$	$\dfrac{1111}{}$
$36663 \div 11 = ?$	$11)\overline{12221}$
$61105 \div 11 = ?$	$\underline{11}$
$24442 \div 11 = ?$	$\overline{12}$
$48884 \div 11 = ?$	$\underline{11}$
$73326 \div 11 = ?$	$\overline{12}$
$85547 \div 11 = ?$	$\underline{11}$
$97768 \div 11 = ?$	$\overline{11}$
$109989 \div 11 = ?$	11

The above series may be followed by:

$$13322 \div 12 = ?$$
$$26664 \div 12 = ?$$
$$39996 \div 12 = ?$$
$$53328 \div 12 = ?$$
$$79992 \div 12 = ?$$
$$93324 \div 12 = ?$$
$$106656 \div 12 = ?$$
$$119988 \div 12 = ?$$

These divisors might well be followed by 13, 14, 15, 16, 18, 19.

Concrete Problems

In the beginning all concrete problems should involve only one of the four fundamental processes. The same kind of problem should be continued until the form is fixed. The solution of written problems may be aided by (1) a pictorial or diagrammatic representation of the problem, (2) an oral interpretation or estimate of the conditions of the problem before the solution is attempted, (3) the simplifying of the situation by using smaller numbers.

Encourage the children to make original problems illustrating the application of the fundamentals. Problems may be based upon the prices of common groceries: beans, sugar, coffee, rice, potatoes, flour, apples, bread; upon articles of clothing, as calico, muslin, linen, silk, collars, shoes, gloves, slippers, buttons; upon playthings; upon farm products; upon household furnishings, and the like. It will stimulate interest if the pupils are encouraged to keep their original problems classified in a permanent note-book.

CHAPTER XII

DENOMINATE NUMBERS

The knowledge of numbers was probably first used in measuring objects. It, perhaps, was almost as difficult to evolve units and standards of measures as it was to evolve a system of numbers. One of the earliest units of measure must have pertained to value. Until some common measure of value was agreed upon it was impossible to compare the possessions of one person with those of another, or to engage in trade upon any equitable basis. The very necessities of exchange where the wealth or property of some men consisted of furs, of others of cattle, of still others of tea, ornaments, tools or weapons, required a medium so that the value of one could be expressed in terms of the other. Occasionally this medium or standard was purely imaginary. John Stuart Mill[1] informs us that certain African tribes calculated the value of things in a sort of money of account, called macutes. One object is worth ten macutes, another twenty, and so on. But mascutes have no real existence; they are imaginary conventional units used for comparing one object with another.

Money the Common Medium of Exchange

But it was not always possible to exchange one object of value for another object of value. Distance must have made it impossible many times for the transfer to have been made on the spot. A medium of exchange, a currency,

[1] *Political Economy*, chapter 5, p. 11.

became a necessity. Usually gold or silver served as money. The reason for this is simple. As soon as people were assured of the necessities of life, they naturally sought after the more precious and the more ornamental things, like gold, silver, and precious stones. These are valuable because they are rare, imperishable and portable. Gold and silver became the basis of exchange rather than jewels because of their greater durability.

Money became both a unit of measure and an equivalent of value. Because it was a measurer and because objects varied in value, divisions and subdivisions of the unit were necessary.

It must be remembered that money was first weighed, not counted. Down to the time of mediaeval England coins were used as weights. This accounts for the English pound as a unit of weight. The Roman pound or libra was determined by a balance called the libra. Thus a libra of gold could be converted in a number of coins, or so many coins equalled a libra of gold. The standard libra varied with different Emperors, but in mediaeval England a pound Troy came to mean the weight of 240 pennies or denarii. The history of the names employed in the various coinage systems would be interesting, but it would take us too far afield. Any adequate account of the endeavors to prevent the corruption of coins, the depreciation of their value, abuses due to dishonest moneyers and exchangers, the attempts to secure a nation-wide or a universal system of coinage, the introduction of a paper equivalent for gold or silver, would require more time and space than we can afford to give.

Troy and Avoirdupois Weights

The Hindus had two original units of weight, the most ancient of which was the Gerah, whose equivalent was twelve grains of barley. The seed of a creeping plant called the rati, was frequently used as the alternative to the barley-grain. The carat, a bean of an Abyssinian tree, which, like the rati-seed was of almost uniform size, was also used as a measure of weight by dealers in precious metals.

The Roman pound consisted of 5204 grains, and each pound was divided into 12 unciae or ounces. Henry III, in the thirteenth century, decreed that "an English penny, called a sterling, sound and without clipping, shall weigh 32 wheat corris in the midst of the ear; and 20 pennies do make an ounce; and 12 ounces one pound." There is no need of additional evidence to show that the primary unit of weight was a grain, sometimes barley, sometimes rati-seed, sometimes wheat.

Historic Relation of Weight and Capacity

The measurements of weight and of capacity have always been closely related. Grains and vegetables are bought in one place by the bushel or the gallon, and in another place by the pound. People experience little difficulty in translating one of these into the other. The expression of this relationship between these two types of measure resulted in a statute during the reign of Henry VII, which statute reads:

"One sterling (or penny) shall be the weight of 32 corris of wheat that grew in the midst of the ear, according to the old laws of the land." The table evolved from the statute was:

32 grains	1 sterling
20 sterlings	1 ounce
12 ounces	1 pound
8 pounds	1 gallon
8 gallons	1 bushel

Here we find the two types of measure, weight and capacity, clearly expressed in a single table.

It was found, however, that all gallons or bushels did not weigh alike. They varied with reference to the kind of material, its dryness, and whether the measures were heaped or struck. Hence it was found to be necessary to state more accurately the meaning of a pound. During the reign of Edward II the term *avoir du pois* was invented to refer to a uniform pound of goods, whether heavy or light. Practice more than legislation resulted in differentiating Troy and Avoirdupois weights; the former came gradually to be used to express a scale of relation for precious metals; the latter for all common materials.

Although a Troy pound is supposed to consist of 5760 grains and an Avoirdupois pound of 7000 grains, so that 7000 grains Troy equal 1 pound Avoirdupois, it is quite generally known that the common grain is no longer used as the base unit. About the middle of the nineteenth century the British Government adopted the length of a second's pendulum oscillation as the invariable unit of length. Of the 39.1393 parts into which it was divided the British yard contains 36. Each of these is the British inch. The distilled water which fills a cubic inch is found to weigh always the same amount, under the same conditions. This weight is divided into 252.422 equal parts; the British pound avoirdupois contains 7000 of these parts.

"Again, it is possible to give the quantity of distilled water which weighs exactly 10 of these pounds avoirdupois. This quantity of water always fills the same space. The space is the 'content' of the British *gallon*, the unit of capacity. With the final precaution that the gallon measure shall be circular at base, not rectangular, and of given diameter, our units of capacity and weight are clearly connected and rest upon the firm basis of the 'oscillation' unit of length."

Step by step the two tables were differentiated. The Troy scale being expressed as follows:

24 grains make 1 pennyweight
20 pennyweights make 1 ounce
12 ounces make 1 pound
Hence 5760 grains make 1 pound Troy
While 7000 grains make 1 pound Avoirdupois

And the Avoirdupois scale:

16 drams make 1 ounce
16 ounces make 1 pound
112 pounds make 1 hundred-weight
20 hundred-weights make 1 ton

The dram or drachm was adopted by Roman physicians instead of the denarius. The most curious part of the Avoirdupois was the use of 112 pounds to equal one hundred-weight. Because the amount of material necessary to make 100 pounds varied with localities and with materials, the value of one hundred-weight varied. Eventually during the reign of Elizabeth it was fixed arbitrarily at 112 pounds.

Length, Surface and Solidity

Units of length are derived mainly from parts of the human body or from movements described by them. The nail, palm, foot, and handsbreadth, clearly refer to parts of the body. An ell represents the distance from the elbow to the tip of the longest finger; the yard, from the armpit to the tip of the longest finger; a span, the distance between the extended tips of the thumb and the middle finger; the pace, two steps; the mile, a thousand paces.

Units of length and of weight had a common origin.

With the Hindus 3 barleycorns placed end to end or 8 placed side by side, measured the length of the longest finger joint, which distance we call an *inch*. It seems quite probable that this notion prevailed with the Arabians, and that the Romans related 12 such parts to the length of the foot. At any rate, by the time of Edward II in England the relationship of various units of length was expressed in this table:

> 3 barleycorns (sound and dry) placed end to
> end make 1 inch
> 12 inches make 1 foot
> 3 feet make 1 yard
> 5½ yards make 1 perch

There is a tradition to the effect that Edward III had the length of his arm registered in a bar of metal, and this was divided first into three equal parts, and each of these into 12 inches. It is clear that different people would use different parts of the table differently. Persons dealing with short measures, the cloth merchants, for example, would become proficient with the yard, foot and inch; masons and carpenters would frequently require a knowledge of both the short and middle distance units, while agriculturalists, soldiers and sailors would more likely employ the longest units.

Square measure was derived directly from linear measure. It really does not require a separate table, as the names of the units and sub-multiple in each instance are the same.

Linear	*Square*
12 in. make 1 ft.	12^2 sq. in. make 1 sq. ft.
3 ft. make 1 yd.	3^2 sq. ft. make 1 sq. yd.

The similarity of terms does not hold beyond this point. Like square measure, cubic measure can easily be deduced from linear measure. It is merely an expression of length in three directions.

Uniformity in Measures

"The subject of weights and measures of a people bears much the same relation to them as does the language of ordinary speech, being assumed and applied in their daily occupations without active thought and resisting changes and reforms even when brought about by the most strenuous efforts and with convincing proof of their desirability' or necessity." Wherever civilization has progressed and society has become more complex, varying and inexact units of measure have been found unsuited to the needs of men. Progress in civilization always necessitates greater uniformity of the units of measure and greater exactness in the units employed.

When the grain was defined to be "the weight of a grain of wheat taken from the middle of the ear and well dried" it was a variable unit and all those units of weight which were defined in terms of the grain were also variable. Our present system of weights and measures has one of the prerequisites of an ideal system—its units are accurately defined and are invariable, but it is a poor system when judged by the other prerequisite—the scale for ascending and descending reduction must be uniform in a given table. The Metric system has both of these prerequisites and is therefore superior to the English system.

The necessity of adopting standards of relations in the United States has been accompanied with a variety of problems. The different states have had no uniform system of weights and measures. To obviate this diffi-

culty there has been an insistent demand for the reforma-
tion of the weights and measures and for the establishment and enforcement of a national system of standards.

The recognition of the English origin and variability
of colonial measures was responsible for the statement in
the Articles of Confederation that: "The United States in
Congress assembled shall also have the sole and exclusive
right and power of regulating the value of alloy and coin
struck by their own authority or by that of the respective
states, and of fixing the standards of weight and measures
throughout the country."

Later the Constitution conferred upon Congress the
power to coin money, regulate the value thereof and of
foreign coin, and fix the standard of weights and measures.[1]

After careful consideration the following were adopted:
The yard of 36 inches, the avoirdupois pound of 7,000
grains, the gallon of 231 cubic inches, and the bushel
of 2,150.42 cubic inches. The standard yard was the
thirty-six inches comprised between the twenty-seventh
and sixty-third inches of a brass bar prepared for the court
survey in London. The avoirdupois pound equals 1.215
pounds troy, which is the relative equivalent between these
weights in England. The units of measures represent the
wine gallon of 231 cubic inches and Winchester bushel.
These units were adopted June 14, 1836, and the secretary
of the treasury was directed to supply a uniform set to the
governor of each state. Having thus provided the means
of uniformity, Congress delegated the responsibility of
enforcing it to the separate states.

The metric system was legalized in this country in 1866,
and is now in daily use in many commercial transactions.
The movement to substitute it for our national system is

[1] Article 1, Section 8.

not powerful enough to warrant the hope of its speedy adoption.

There is, however, a growing sentiment in favor of greater simplicity and uniformity as to the method involved in selling those commodities that are measured by the quantity or bulk. This sentiment favors the selling of such products and articles by weight. By the law of 1866 a bushel of wheat weighs sixty pounds, of corn or rye fifty-six pounds, of barley forty-eight pounds, of oats thirty-two pounds, of peas sixty pounds, and of buckwheat forty-two pounds. Duties imposed upon these products are based upon weights rather than bulk measure. Of course it would be false to assume that each of these weights equals a volume equivalent to the standard bushel 2150.42 cubic inches. These weights simply represent the averages obtained for these products in different localities for a numbers of years.

The Teaching of Denominate Numbers

During the Middle Ages the subject of denominate numbers was of great importance to anyone who expected to engage in commerce. Most of the cities and states had their own systems of weights and measures and it was not uncommon for the systems of two neighboring cities to be quite unlike. A knowledge of the most important systems was necessary in order that the reductions from one to another could be readily made. The arithmetics of the period contained numerous tables, and the lack of decimal fractions necessitated compound numbers of several denominations.

Teachers who have taught some years are familiar with the fact that much of the material formerly included in tables of measures has been eliminated. The movement for greater simplicity in arithmetic is as evident in denominate numbers as elsewhere. The modern tendency is to teach only those tables that are generally employed by the majority of people and to leave the technical tables of the physician, the druggist, and the jeweler to be learned by those who will have occasion to use them. The newer arithmetics have eliminated in addition to Troy and Apothecaries' and Surveyor's measures, the following:

$$12 \text{ things} = 1 \text{ dozen (1 doz.)}$$
$$12 \text{ dozen} = 1 \text{ gross (1 gro.)}$$
$$12 \text{ gross} = 1 \text{ great gross (g. gross)}$$
$$20 \text{ things} = 1 \text{ score}$$
$$24 \text{ sheets of paper} = 1 \text{ quire}$$

20 quires

or   = 1 ream

480 sheets

$$16 \text{ cubic feet} = 1 \text{ cord foot (cd. ft.)}$$

$$\left. \begin{array}{l} 128 \text{ cubic feet} \\ 8 \text{ cord feet} \end{array} \right\} = k \text{ cord (cd.)}$$

$$6 \text{ feet} = 1 \text{ fathom}$$
$$6086.7 \text{ feet} = 1 \text{ knot}$$
$$3 \text{ knots} = 1 \text{ league}$$

It is desirable that the tables should be collected instead of appearing in a random fashion throughout the text.

There are two things of paramount importance to be considered in the teaching of denominate numbers: one is the reduction or changing of numbers either to larger or to smaller units, and the other is the automatic mastery of certain simple combinations that should be known at sight. In teaching the reduction of denominate numbers both whole numbers and fractional measures should be used.

Much of the work in denominate numbers should be oral.

Many problems to be solved at sight should be given. In every instance they should be related to the daily trans- .actions of business life. Long written problems in the addition, subtraction, multiplication, and division of denominate numbers should be omitted. Problems involving reduction through more than three denominations are seldom used in the business world. Children should be made familiar with the simple and necessary facts of linear measure by drawing, measuring, building, counting, and grouping; of the similar facts of liquid and dry measure by measuring liquids and a few common products, like potatoes or tomatoes, and by applying the measures to products bought for home use. A knowledge of the other units of measure should be acquired and applied in the same way. As far as possible children should have the actual measures presented to their senses. It is easy to make the mistake of talking about these things without children's actually having a knowledge of them.

Addition, subtraction, multiplication, and division of denominate numbers are based upon the same principles as those previously developed for integers.

CHAPTER XIII

COMMON FRACTIONS

Common fractions have been regarded as one of the most difficult topics in arithmetic. Many young teachers have hesitatingly approached the subject and have grown discouraged as they attempted to teach children who were blindly floundering their way "through fractions." The difficulty has been magnified because fractions have been taught frequently as if they were a new system of notation, because the difficult character of the subject has supplied the occasion for numerous short lectures by teachers to pupils, and because emphasis has been placed upon the automatic mastery of the mechanical phases of fractional relations, while the rational processes have been ignored. To have it constantly dinned into your ears that here is a new and very hard kind of work, tends to increase rather than to lighten the burden. Small wonder that teachers get discouraged and pupils perplexed!

The last of the reasons assigned above has been responsible for a most unfortunate result. It is a survival of the mediaeval tendency to substitute too early counting and abstract work for contact with objects. Whenever this is done the number sense becomes as obtuse as it was before Pestalozzi reintroduced object teaching. The adult mind is prone to break fields up into logical classifications and to insist upon an almost verbal mastery of them, neglecting at the same time those situations that gave birth to the fact and in which the fact finds its present significance. Children do not see the world broken up in this way. They grasp general truths through contact with things. They must have objects to weigh, to measure, to evaluate, to manipulate. It is through the handling of them that the interpretative powers of children receive their initial training.

Stages in Teaching Fractions

There are three clearly recognizable and well-defined stages in the teaching of fractions. Their order is (1) the meaning, (2) the recognition, and (3) the manipulation of fractions. The language used in fractions does not always convey the intended meaning to the pupil. The difficulty is not due so much to language as to the vague idea the average pupil has of fractions. The language of fractions must be so phrased that attention is fixed upon the notion that fractions are concrete numbers instead of abstractions ingeniously devised to be manipulated. The limited knowledge of fractions children have when they come to school has been acquired from the handling of objects common to their play and the home. They do not think $\frac{1}{2}$ or $\frac{1}{3}$ as symbolically represented, but as referring to particular things. Their idea is expressed by the separation of a unit or of groups into equal parts.

Instruction must begin with the knowledge children have. In fractions, exercises may be given in having a child divide some object, say an apple, into two or three equal parts. The child may then count the parts and tell how many he has. When this operation has been repeated enough times, attention may be directed to the units involved. $\frac{2}{3}$ of an apple is a fraction. The whole apple is the original unit, and the one-third of an apple is a fractional unit. The two indicates the *number* of pieces of the apple taken. The 3 indicates the *kind* of *units* taken.

Thus there are at least three steps in acquiring a fractional concept: (1) a unit divided into equal parts, (2) a comparison of one or more of these parts with the original unit, (3) a group of these parts thought together. The symbols used to express fractional ideas need not be taught until later.

The concepts for one-half, one-third, and one-fourth should be taught at the time children are receiving their early instruction in integers. The constructive work of the primary grades is admirably adapted for this purpose. Folding exercises develop ideas of two, three, and four equal parts of a unit. They are excellent devices for developing the habit of self-criticism and of training in accuracy. There are other means, incidental in character, for inculcating the simpler fractional concepts. Fractions appear in separating classes into divisions, sessions, into periods, in the forming of lines of march, the use of rulers, in the distribution of materials. Opportunities to use these will not be overlooked by the industrious and thoughtful teacher.

A fact not always made clear to beginners is that a fraction is as truly a unit as an integer. There are both integral units and fractional units. The fraction one-half may be separated into fourths, and the fourth into sixteenths. Any part may be considered as a unit and may be divided into equal parts in the same way that the integer was separated to get the part. This idea is easily developed by the use of objects, and pictures.

Definition of a Fraction

A commonly accepted but faulty definition of a fraction is that it is one of the equal parts of a unit. That this definition is incomplete and unsatisfactory is seen when we apply it to $\frac{2}{3}$ or $\frac{3}{4}$. These are more than "one" of the equal parts of a unit. No definition is adequate that climinates the majority of the cases that should come under it. A fraction may be correctly defined as *one or more* of the equal parts of a unit. This definition covers all the cases; it includes the separation of unity into equal parts, or of groups of integers, for example, 10, which may be divided into still smaller but equal groups.

Devices in Teaching Fractions

Although we have advocated the teaching of fractions by means of objects, we are not unmindful of the fact that the amount of object teaching possible and desirable varies with the age and intellectual maturity of the class. No hard and fast rule should be followed, for classes differ widely as to their ability and attainment. Object work should be followed by pictures, lines, circles, squares, and rectangles. Fractional notions may be represented by the line as follows:

$\frac{1}{2}$				$\frac{1}{2}$			
$\frac{1}{4}$		$\frac{1}{4}$		$\frac{1}{4}$		$\frac{1}{4}$	
$\frac{1}{8}$	$\frac{1}{8}$	$\frac{1}{8}$	$\frac{1}{8}$	$\frac{1}{8}$	$\frac{1}{8}$	$\frac{1}{8}$	$\frac{1}{8}$

changing fractional forms may be shown by the circle.
How many fourths in one-half?
How many eighths in one-fourth?
How many eighths in one-half.

Needed Eliminations

It is unnecessary to extend illustrations of the possible uses of pictures and drawings. In one of the newer primary arithmetics more than two hundred pictures of lines, circles and squares are used in various ways to teach fractions. These are not used as extensively as they should be.

Teachers should confine themselves largely to those fractions that are used in every day business life. Long fractions, as $\frac{7890570110}{15781140200}$ and exceedingly complex fractions, as

$$\frac{\dfrac{1}{3}+\dfrac{4}{7}+\dfrac{21}{32}}{\dfrac{2}{5}+\dfrac{23}{56}+\dfrac{3}{5}} \quad \text{or} \quad \frac{4\dfrac{1}{20}+8\dfrac{2}{15}}{4\dfrac{1}{15}+8\dfrac{1}{10}}$$

and other unbusiness-like forms should be omitted. In earlier days fractions of this sort were believed to be of peculiar value in sharpening the wits. of people. This afforded sanction enough for their use in the past, but such a reason is discredited to-day.

The History of Fractions

The history of the evolution of fractions shows that modifications come about slowly. The oldest known mathematical records indicate that fractions were extensively used at an early date. The systems, however, were cumbersome and the methods peculiar and complicated. This partly accounts for the tradition that fractions are difficult. The work of Ahmes, which dates back to 2000 or 3000 B. C., devotes considerable space to fractions. Most of his fractions had unity for a numerator. He expressed his fractions by writing the denominator and placing over it either a dot or a symbol called *ro*. Whenever a fractional value occurred that could not be thus expressed, it was written as the sum of two or more unit fractions. $\frac{2}{9}$ was regarded as $\frac{1}{16}+\frac{1}{18}$, $\frac{2}{97}$ as $\frac{1}{56}+\frac{1}{679}+\frac{1}{776}$. In the case of $\frac{2}{3}$ an exception was made and a symbol was used to represent it. The Babylonians used sexagesimal fractions; that is, fractions having a constant denominator of 60. They wrote the numerator only, placing it a little to the right of the ordinary position. The Greeks at first used *unit* fractions and indicated them by simply writing the denominator with a double accent. Later they used ordinary fractions, writing the numerator once with a single accent and the denominator twice with a double accent. They would have written $\frac{9}{15}$ as 9′ 15″ 15″. The Romans used duodecimal fractions; that is, fractions having 12 for the denominator. The inch and the ancient ounce were outgrowths of this. The Hindus used the general fractions and indicated them by writing the numerator above the denominator without any line separating them. Thus $\frac{2}{3}$ was written $\frac{2}{3}$. The Arabs introduced the separating line. In the oldest Hindu arithmetic, the Liliwati, the multiplication of fractions is indicated by writing the fractions consecutively without any symbol between them. Addition of a fraction to an integer was shown by writing the fraction beneath the integer, and subtraction was shown in the same way, with a dot prefixed to the fraction. Thus $4+\frac{2}{3}$ was written $\overset{4}{\frac{2}{3}}$; and $4-\frac{3}{4}$, as $\overset{4}{\frac{3}{4}}$. The problem, "Tell me, dear woman, quickly, how much a fifth, a quarter, a half, and a sixth make when added together," appears solved:

1	1	1	1	67
5	4	2	6	60

That the subject of common fractions was a difficult one for the race is indicated by the fact that unit fractions were used for thousands of years. During the Middle Ages the term "common" was used to distinguish these fractions from sexagesimal fractions. Common fractions are called "vulgar" fractions in England.

Since the Arab invasion of Europe there has been almost
no change in the formal representation of fractions. Many
of the rules found in Liliwati of the Hindus remain practically unchanged to-day. But fractions have increased
vastly in importance; they are a common possession of the
common people; they are constantly needed in our every-
day transactions. Text-books give far more space to them
than formerly, the problems are more real and the instruction more skillful.

Interpreting Fractions

One of the chief difficulties, however, yet remains: the
difficulty experienced by children in interpreting clearly
and quickly the perplexing forms and terms used. It is
to this end that we so insistently urge that children learn
to objectify common fractional quantities, halves, fourths,
sixths, etc., only after they have first learned them in
connection with actual objects. When the pupils have
acquired skill in recognizing these, they should be required
to re-apply them to objects and to pictorial representations
of objects. Drawings, because they are subject to a wider
application than objects, constitute the natural transitional
material between objects and abstract work. A knowledge
of the simpler fractional parts may be facilitated by ar-
ranging them into related groups:

1. The whole, halves, fourths.
2. The whole, halves, fourths, eighths.
3. The whole, thirds, sixths, twelfths.
4. The whole, thirds, fourths, and twelfths.

(Suggested by the Indianapolis Course of Study.)

"In each of the figures show the whole; a half; a fourth.
How many halves in the whole? Show them. How many
fourths in the whole? Show them. How many fourths in

the half? Show them. Compare in as many ways as possible, pointing out each fractional part named: the whole and $\frac{1}{2}$; the whole and $\frac{1}{4}$; $\frac{1}{2}$ and $\frac{1}{4}$.

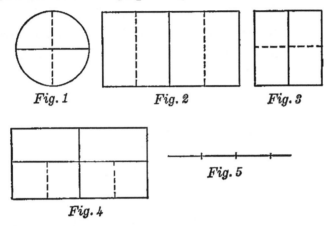

Fig. 1   Fig. 2   Fig. 3

Fig. 5

Fig. 4

The Fundamentals of Fractions

After the pupils thoroughly understand and appreciate what a fraction is and have learned to write fractional quantities, they should be taught to reduce, to add, subtract, multiply and divide them. It is best to begin these operations with fractions that appear as products in the multiplication table. Next in importance to the fundamental concept that a fraction is one or more of the equal parts of a unit, is the idea that fractions may be changed to equivalent fractions by multiplying both terms by the same number. It requires considerable skill to make clear the notion that there may be a change of form without a change of value. If we take the fraction $\frac{3}{4}$ and multiply its terms by 2, we get $\frac{6}{8}$. Now, the $\frac{1}{8}$ of a foot is an inch and a half, and therefore $\frac{6}{8}$ of a foot is 6 inches and 6 half inches, or 9 inches; but $\frac{3}{4}$ of a foot is 9 inches, therefore $\frac{3}{4}$ of a foot equals $\frac{6}{8}$ of a foot. The teacher must show that this principle is true whatever the unit of measure may be, or whatever the fraction of that unit. If this is well done the pupils have learned the "secret" of reducing fractions to lower or raising them to higher terms. They then understand that the same fractional value may have many different names. The table listed below shows this fact, and it may be used also to find a common denominator for different fractions.

	4ths	6ths	8ths	9ths	10ths	12ths	14ths	15ths	16ths	18ths	20ths	21sts	22ds	24ths
$\frac{1}{2}$	$\frac{2}{4}$	$\frac{3}{6}$	$\frac{4}{8}$	$\cdot\cdot$	$\frac{5}{10}$	$\frac{6}{12}$	$\frac{7}{14}$	$\cdot\cdot$	$\frac{8}{16}$	$\frac{9}{18}$	$\frac{10}{20}$	$\cdot\cdot$	$\frac{11}{22}$	$\frac{12}{24}$
$\frac{1}{3}$	$\cdot\cdot$	$\frac{2}{6}$	$\cdot\cdot$	$\frac{3}{9}$	$\cdot\cdot$	$\frac{4}{12}$	$\cdot\cdot$	$\frac{5}{15}$	$\cdot\cdot$	$\frac{6}{18}$	$\cdot\cdot$	$\frac{7}{21}$	$\cdot\cdot$	$\frac{8}{24}$
$\frac{1}{4}$	$\frac{1}{4}$	$\cdot\cdot$	$\frac{2}{8}$	$\cdot\cdot$	$\cdot\cdot$	$\frac{3}{12}$	$\cdot\cdot$	$\cdot\cdot$	$\frac{4}{16}$	$\cdot\cdot$	$\frac{5}{20}$	$\cdot\cdot$	$\cdot\cdot$	$\frac{6}{24}$
$\frac{1}{5}$	$\cdot\cdot$	$\cdot\cdot$	$\cdot\cdot$	$\cdot\cdot$	$\frac{2}{10}$	$\cdot\cdot$	$\cdot\cdot$	$\frac{3}{15}$	$\cdot\cdot$	$\cdot\cdot$	$\frac{4}{20}$	$\cdot\cdot$	$\cdot\cdot$	$\cdot\cdot$
$\frac{1}{6}$	$\cdot\cdot$	$\frac{1}{6}$	$\cdot\cdot$	$\cdot\cdot$	$\cdot\cdot$	$\frac{2}{12}$	$\cdot\cdot$	$\cdot\cdot$	$\cdot\cdot$	$\frac{3}{18}$	$\cdot\cdot$	$\cdot\cdot$	$\cdot\cdot$	$\frac{4}{24}$
$\frac{1}{7}$	$\cdot\cdot$	$\cdot\cdot$	$\cdot\cdot$	$\cdot\cdot$	$\cdot\cdot$	$\cdot\cdot$	$\frac{2}{14}$	$\cdot\cdot$	$\cdot\cdot$	$\cdot\cdot$	$\cdot\cdot$	$\frac{3}{21}$	$\cdot\cdot$	$\cdot\cdot$
$\frac{1}{8}$	$\cdot\cdot$	$\cdot\cdot$	$\frac{1}{8}$	$\cdot\cdot$	$\cdot\cdot$	$\cdot\cdot$	$\cdot\cdot$	$\cdot\cdot$	$\frac{2}{16}$	$\cdot\cdot$	$\cdot\cdot$	$\cdot\cdot$	$\cdot\cdot$	$\frac{3}{24}$
$\frac{1}{9}$	$\cdot\cdot$	$\cdot\cdot$	$\cdot\cdot$	$\frac{1}{9}$	$\cdot\cdot$	$\cdot\cdot$	$\cdot\cdot$	$\cdot\cdot$	$\cdot\cdot$	$\frac{2}{18}$	$\cdot\cdot$	$\cdot\cdot$	$\cdot\cdot$	$\cdot\cdot$
$\frac{1}{10}$	$\cdot\cdot$	$\cdot\cdot$	$\cdot\cdot$	$\cdot\cdot$	$\frac{1}{10}$	$\cdot\cdot$	$\cdot\cdot$	$\cdot\cdot$	$\cdot\cdot$	$\cdot\cdot$	$\frac{2}{20}$	$\cdot\cdot$	$\cdot\cdot$	$\cdot\cdot$
$\frac{1}{11}$	$\cdot\cdot$	$\cdot\cdot$	$\cdot\cdot$	$\cdot\cdot$	$\cdot\cdot$	$\cdot\cdot$	$\cdot\cdot$	$\cdot\cdot$	$\cdot\cdot$	$\cdot\cdot$	$\cdot\cdot$	$\cdot\cdot$	$\frac{2}{22}$	$\cdot\cdot$
$\frac{1}{12}$	$\cdot\cdot$	$\cdot\cdot$	$\cdot\cdot$	$\cdot\cdot$	$\cdot\cdot$	$\frac{1}{12}$	$\cdot\cdot$	$\cdot\cdot$	$\cdot\cdot$	$\cdot\cdot$	$\cdot\cdot$	$\cdot\cdot$	$\cdot\cdot$	$\frac{2}{24}$

The preceding table was not given to be memorized. There are, however, some common fractions that occur so frequently in business that every one should know them. If the preceding equalities have been clearly taught and apprehended, pupils will appreciate the desirability of learning equivalent decimal forms and equivalent per cents.

Common Fractions		Decimal Fractions		Parts of $1		Per cents
$\frac{1}{2}$	=	.5	=	$.50	=	50%
$\frac{1}{4}$	=	.25	=	$.25	=	25%
$\frac{1}{3}$	=	.33$\frac{1}{3}$	=	$.33$\frac{1}{3}$	=	33$\frac{1}{3}$%
$\frac{1}{5}$	=	.2	=	$.20	=	20%
$\frac{1}{6}$	=	.16$\frac{2}{3}$	=	$.16$\frac{2}{3}$	=	16$\frac{2}{3}$%
$\frac{1}{8}$	=	.12$\frac{1}{2}$	=	$.12$\frac{1}{2}$	=	12$\frac{1}{2}$%
$\frac{1}{10}$	=	.1	=	$.10	=	10%
$\frac{3}{4}$	=	.75	=	$.75	=	75%
$\frac{2}{3}$	=	.66$\frac{2}{3}$	=	$.66$\frac{2}{3}$	=	66$\frac{2}{3}$%
$\frac{2}{5}$	=	.4	=	$.40	=	40%
$\frac{3}{5}$	=	.6	=	$.60	=	60%
$\frac{4}{5}$	=	.8	=	$.80	=	80%
$\frac{5}{6}$	=	.83$\frac{1}{3}$	=	$.83$\frac{1}{3}$	=	83$\frac{1}{3}$%
$\frac{3}{8}$	=	.37$\frac{1}{2}$	=	$.37$\frac{1}{2}$	=	37$\frac{1}{2}$%
$\frac{5}{8}$	=	.62$\frac{1}{2}$	=	$.62$\frac{1}{2}$	=	62$\frac{1}{2}$%
$\frac{7}{8}$	=	.87$\frac{1}{2}$	=	$.87$\frac{1}{2}$	=	87$\frac{1}{2}$%

Reduction of Fractions

The "reduction of fractions to equivalent fractions having a common denominator" is necessary before unlike fractions can be added or subtracted. We can add $\frac{4}{7}$, $\frac{5}{7}$, $\frac{6}{7}$, as they are all of the same denomination—sevenths. We can easily see that to add $\frac{2}{3}$ and $\frac{1}{6}$, we have only to change $\frac{2}{3}$ to sixths. Similarly many fractions can be changed to equivalent fractions by inspection. To find the value of $\frac{3}{4} + \frac{5}{8} + \frac{7}{8}$ is slightly more difficult, but not essentially different in principle. In addition of dissimilar fractions the student is confronted with the problem of discoverng some common number or multiple to which all the different fractions may be related. A simple way of finding this common multiple is by finding the product of the prime factors of the denominators, using each factor the greatest number of times it appears in any number. In the example above we get 24. Thus, if both the terms of $\frac{3}{4}$ be multiplied by 6 we get $\frac{18}{24}$.   Similarly the other fractions may be changed to equivalent fractions having the common denominator 24, and the addition performed.

What has been said about addition applies with equal force to the subtraction of fractions. The order of difficulty is the same; first teach the subtraction of similar fractions; then of dissimilar fractions, and finally of mixed numbers. Addition and subtraction may be taught simultaneously. Pupils in the sixth grade should be able to state instantly the result of such examples as: $\frac{1}{2}+\frac{1}{3}$; $\frac{1}{5}+\frac{1}{4}$; $\frac{1}{2}-\frac{1}{3}$ and $\frac{1}{5}-\frac{1}{8}$.

Multiplication and Division

In teaching multiplication and its reverse process, division, the procedure should be (1) multiplying or dividing a fraction by a whole number, (2) a whole number by a fraction, (3) a fraction by a fraction, (4) a mixed number by a whole number, (5) a mixed number by a mixed number.

The principles underlying these operations are few in number and simple in nature. They usually appear as rules. But they should be so presented that self-discovery on the part of the pupils is inevitable. To some of these principles we have already alluded; for example, the principle that fractions may be changed in form without altering their value by multiplying or dividing both terms by the same number. A second equally important truth is that only like fractions can be added or subtracted. The third and fourth principles refer to the ways for increasing the value of a fraction. Multiplying the numerator by an integer multiplies the value of the fraction by making more parts in the fraction, and dividing the denominator by an integer multiplies the value of the fraction by making the size of each part larger. The fifth and sixth principles refer to the ways for diminishing a fraction. Dividing the numerator by an integer divides the value of the fraction by decreasing the number of parts, and multiplying the denominator by an integer divides the values of the fraction by making the parts smaller.

It will be noted that the expressions multiplication and division, as applied to fractions, are extensions of the ordinary meanings of those terms, for the former, in its original meaning, implies increase, and the latter decrease; but when two proper fractions are multiplied together the product is less than either of the factors, and when one proper fraction is divided by another, the quotient is greater than either the dividend or the divisor.

There are several explanations for the inversion of the divisor in the division of fractions.

1. Inverting the divisor gives the same result that is obtained by reducing both fractions to their least common denominator and dividing the numerator of the dividend by the numerator of the divisor.

$$\tfrac{2}{5} \div \tfrac{3}{4} = \tfrac{8}{20} \div \tfrac{15}{20} = \tfrac{8}{15}.$$

The same idea is involved here as in the division of \$8 by \$15; the result is $\tfrac{8}{15}$.

But $\tfrac{2}{5} \times \tfrac{4}{3} = \tfrac{8}{15}$; therefore $\tfrac{2}{5} \div \tfrac{3}{4} = \tfrac{2}{5} \times \tfrac{4}{3} = \tfrac{8}{15}$.

2. If the divisor were 3, then the quotient would be $\tfrac{1}{3}$ of $\tfrac{2}{5}$, but the divisor is $\tfrac{1}{4}$ of 3; hence the quotient is $4 \times \tfrac{1}{3}$ of $\tfrac{2}{5}$, which is $\tfrac{4}{3} \times \tfrac{2}{5}$, or $\tfrac{8}{15}$.

3. $\tfrac{2}{5} \div \tfrac{3}{4} = $ Quotient.

Therefore $\frac{3}{4}$ Quotient $=\frac{2}{5}$.

$\frac{1}{4}$ Quotient $=\frac{1}{3}$ of $\frac{2}{5}=\frac{2}{15}$.

$\frac{4}{4}$ Quotient $=4\times\frac{2}{15}=\frac{8}{15}$.

4. Reasoning from known to unknown.

$\frac{2}{5}\div\frac{3}{4}=?$

$\frac{3}{4}$ contains $\frac{3}{4}$, 1 time.

$\frac{1}{4}$ contains $\frac{3}{4}$, $\frac{1}{3}$ time.

1 contains $\frac{3}{4}$, $\frac{4}{3}$ times.

$\frac{2}{5}$ contains $\frac{3}{4}$, $\frac{2}{5}\times\frac{4}{3}$ times;

therefore $\frac{2}{5}\div\frac{3}{4}=\frac{8}{15}$.

This is one of the most difficult for children to understand.

The mastery of common fractions is no longer attempted in any one grade. Some attention is given to them in all of the grades, but most of the formal treatment of the subject is found in the fifth grade. Complex fractions are omitted in most treatments of the subject. It is wise to teach the operations with only as much theory as seems necessary to a clear understanding of the subject, and pupils should not be required to memorize and repeat the explanations. It is necessary that teachers understand thoroughly the reasoning underlying the various processes, but it is not desirable that we attempt to force this mature reasoning upon the pupils. Results should be gauged by the knowledge and skill attained, not by the number of pages covered.

CHAPTER XIV

DECIMAL FRACTIONS

Origin of Decimal Fractions

"The invention of decimal fractions, like the invention of the Arabic scale, was one of the happy strokes of genius."[1] The decimal fraction is a comparatively recent product of man's ingenuity. As Napier says: "It is a surprising fact that decimals, so simple and convenient, were not invented until after so much had been attempted in physical research and numbers had been so deeply pondered." Decimals were unknown to the Greeks and the Romans, whose number systems did not utilize place value in the sense in which we use the term.

It is not known in what way decimal fractions actually originated. There are two theories which account for their origin. The advocates of the first theory believe that they were devised as a shorthand method for expressing certain common fractions. They assert that since the base of our number system is ten, someone devised or invented decimal fractions as a transition from common fractions. If this theory is correct, and if we follow the historical development of the subject in our teaching, the intimate relationship to common fractions should be emphasized. This is the method followed by some text-books.

Those who advocate the second theory maintain that the subject arose directly from the decimal scale for in-

---

[1] Brooks, "Philosophy of Arithmetic," p. 443.

tegral numbers. Since numbers diminish in value from left to right in a ten-fold ratio, because of the place value, it would be natural to extend the number scale to the right of the unit. Such an extension of the scale would give rise to the decimal fraction. Consider the number 111. The figure at the right indicates, by virtue of place value, one-tenth of the value indicated by the figure immediately to its left, and 0.01 of that indicated by the second figure to its left. If this same scale is extended to the right of the units figure we have the decimal fraction. If decimals originated in this manner the subject is intimately related to integers and only indirectly related to common fractions. A rational explanation of the processes involved in decimal fraction may be based upon either of these theories. De Morgan asserts that a table of compound interest first suggested decimal fractions.

Decimal fractions were introduced so gradually that their origin cannot be ascribed to any one person. About 1585 a book was written by Stevin in which he attempted to show their great practical value. He says, "It is certain that if the nature of man remains in the future as it is now, he will not always neglect so great an advantage." The world is not always quick to take advantage of a great improvement, as was evidenced by the tardiness in the adoption of the Gregorian calendar and as is evidenced to-day by the opposition to the metric system. It was not until a century and a half after the time of Stevin that decimal fractions were generally taught. Some of the common fractions of the fifteenth century had become so unwieldy that a change was imperative. One author used the fraction $34\frac{99873}{81443}$, and another required the square root of $2520\frac{24267352095}{319794774016}$. The time was ripe for a new notation in fractions.

Symbolism of Decimal Fractions

Decimal fractions appeared in a number of interesting forms before the decimal point was used. It is instructive for teachers to know that the point is not necessary in writing decimal fractions. Fractions had been used for many years before the decimal point appeared. Stevin, who was the first man to write a valuable treatise on the subject, seemed to appreciate the significance of the new fractions, but his system lacked a suitable notation. He did not use the decimal point. If he wished to express 5.992 he wrote it $^{0}_{5}{}^{1}_{9}{}^{2}_{9}{}^{3}_{2}$. He used a zero to indicate the units figure and placed a "1" over the first decimal figure. Various devices were used by other writers to express decimal fractions. The following methods were in actual use for such a number as 5.992:

$$\overset{0}{5}\ \overset{1}{9}\ \overset{11}{9}\ \overset{111}{2} \qquad 5992\overset{111}{} \qquad 5992^{3} \qquad 5|992 \qquad 5\ 992, \qquad 5|\overline{992}$$

Pellos (1492) unwittingly made use of the decimal point. This is the first instance known in which it was used in a printed work. The point was also used by Jobst Bürgi in a manuscript of 1592, and by Pitiscus in some trigonometric tables in 1612. Kepler in 1616 used both the decimal point and the parenthesis to separate the integer and the decimal fraction.

The awkward symbolism for decimal fractions gradually disappeared and either the point or the comma came to be generally used. The symbolism is not settled even to-day. In the United States three and twenty-five hundredths is expressed as 3.25. In England it is expressed as 3·25, while in Germany, France, and Italy it is 3,25 or 3_{25}. A mere space to indicate the separation of the integral and decimal units is not uncommon in print. A zero is sometimes written to the left of the decimal point to call attention to the point more quickly. Thus ·7 may be written 0.7.

It is important that the teacher should see that by any one of the above devices a decimal fraction may be expressed as truly as if the decimal point were used. All of the devices have a common aim, and that is to indicate where the integral part of the number ends and the decimal part begins. Any symbol or combination of symbols that will indicate this may be used instead of the decimal point. Since the purpose served by the symbols just noted is to separate the two kinds of units,—integral and decimal,— the name *separatrix* is given to any symbol that serves this purpose. The point is but one of a large number of symbols that might be used. It appeals to us as the best separatrix, but it was not extensively used until the first quarter of the eighteenth century. "Simple as they now appear, the development of decimal fractions was too great for any one mind or age. The idea of their use gradually dawned upon the mind, one mathematician taking up what another had timidly begun, added an idea or two, until the whole subject was at length fully conceived and developed."[1]

Reading and Writing Decimal Fractions

After the conception of the decimal fraction has been made clear, considerable practice should be given in the reading and writing of decimals. Teachers will find this work greatly facilitated if they require the pupils to memorize the orders of the decimal scale and the position of each order with reference to the decimal point. For example, the pupil should know that thousandths is the third decimal order, millionths the sixth, hundredthousandths the fifth, etc. If asked to write four hundred twenty-three millionths, the pupil should at once think

[1] Brooks, "Philosophy of Arithmetic."

that since millionths is the sixth order, and since there are three digits in four hundred twenty-three, he must insert three zeroes; he therefore begins at the decimal point and writes three zeroes, then the figures 4, 2, and 3. There is an advantage in being able to write decimal fractions from left to right just as we write integers. The pupil who writes decimal fractions from left to right does not need to count the digits after he has written the expression in order to insure the correctness of his work. We do not count the digits after we have written an integer to insure correctness, and there is no reason why we should do so in writing decimal fractions. Such work as the following is of no little value: How would you write twenty-four ten-thousandths? *Answer.* Decimal point, two zeroes, 2, 4. When pupils are required to write decimal fractions from dictation, the enunciation should be very distinct and the pupil should image the decimal completely before writing it; he should then write it continuously from left to right.

To read a pure decimal: Read as in whole numbers, then state the name of the decimal order of the figure at the right. Thus .047 is read forty-seven thousandths.

To read mixed decimals: Read the integral part, then the decimal part, joining the two parts by *and*. Thus 253.047 is read two hundred fifty-three *and* forty-seven thousandths.

Mixed decimals may also be read as improper fractions. Thus 2.47 may be read as two hundred forty-seven hundredths. It is a good exercise to read decimal fractions as tenths, hundredths, etc. Thus 253.047 read as tenths is 2530.47 tenths; as hundredths it is 25304.7 hundredths; as tens it is 25.3047 tens.

Pure or mixed decimals may also be read by stating the order which each digit occupies. Thus 253.047 may be read as two hundreds, five tens, three units, four hundredths, seven thousandths. Hundredths is often read as per cent. Thus .72 may be read 72 per cent. 3.75 may be read three hundred seventy-five per cent.

There are certain types of decimal fractions that are frequently read incorrectly. Illustrations of these types follow:

2400.0006 should be read, twenty-four hundred *and* six ten-thousandths.

.2406 should be read, twenty-four hundred six ten-thousandths.

$.0\frac{2}{3}$ should be read, two-thirds of a tenth, or two-thirds tenths.

$.00\frac{2}{3}$ should be read, two-thirds of a hundredth, or two-thirds hundredths.

$2.0\frac{2}{3}$ should be read, two *and* two-thirds of a tenth.

$0.2\frac{2}{3}$ should be read, two and two-thirds tenths.

$2.2\frac{2}{3}$ should be read, two *and* two and two-thirds tenths.

Attention should be directed to the fact that in reading mixed decimals the word *and* should be used after reading the integral part.

The expression $.\frac{2}{5}$ is sometimes incorrectly used for $.0\frac{2}{5}$. $.0\frac{2}{5}$ is two-fifths of a tenth, but the expression $.\frac{2}{5}$ has no meaning whatsoever in our system of notation. This may be shown by the following: $\frac{2}{5} = .40$; two-fifths of a tenth equals $\frac{2}{5}$ of $\frac{1}{10} = \frac{2}{50} = .04 = .0\frac{2}{5}$. It therefore appears that $.\frac{2}{5}$ is neither $\frac{2}{5}$ of a unit nor $\frac{2}{5}$ of a tenth.

Comparative Values

Pupils should be able to recognize comparative value in decimal fractions. Ask the pupils how much length is represented by each figure in .546 miles? How much money is represented by each figure in $.394? Pupils should realize that 0.1 is greater than .098 and that .01 is greater than .00998. Ask the pupil to state the decimal fraction of two digits that is nearest in value to .5743 or to .03875. It is not wise to emphasize the decimal orders beyond the sixth, and most of the emphasis should be placed upon the first four orders.

Annexing Zeroes

The effect of annexing zeroes to the right of a decimal fraction should be understood. Ask the pupil to write .14; this is the same as .1+.04. If he should write a 3 to the right of the 4, the expression would be .143. How much has been added? Evidently .003 has been added. If a 2 had been written instead of a 3, evidently .002 would have been added. If a zero had been written, evidently no thousands would have been added, or, in other words, the value of the original expression would not have been changed. By numerous illustration such as this the truth should be taught.

The effect of annexing zeroes to the right of a decimal fraction may be explained also by showing that the annexing of zeroes is equivalent to multiplying both numerator and denominator by the same power of 10. For example, $.14 = \frac{14}{100} = \frac{140}{1000} = \frac{1400}{10000} = .1400$.

The thing to be chiefly emphasized in the teaching of decimal fractions is the four fundamental operations.

Addition and Subtraction

Addition and subtraction offer no difficulties except those previously met in the same processes with integers. Care should be taken that units of the same order are placed in the same vertical column so that the decimal points stand in a vertical line. The decimal point in the result is then placed directly under the line of decimal points.

Illustration:

$$
\begin{array}{r}
17.326 \\
0.0437 \\
9.801 \\
0.42 \\
\hline
27\ 5907
\end{array}
$$

Multiplication and Division

There are two methods of explaining the location of the decimal point in multiplication and division of decimal fractions. These two methods are based upon the two conceptions of the origin of the fractions. By one of these methods the position of the decimal point in the result is determined by principles of common fractions; by the other it is determined from the pure decimal conception of the number system.

By the first method the decimal fractions are reduced to common fractions and the multiplication, or division, is then performed. The result is then written as a decimal fraction. Multiplication will first be considered.

Illustrations:

$$.3 \times .15 = \tfrac{3}{10} \times \tfrac{15}{100} = \tfrac{45}{1000} = .045$$
$$.04 \times .007 = \tfrac{4}{100} \times \tfrac{7}{1000} = \tfrac{28}{100000} = .00028$$
$$4.27 \times .005 = \tfrac{427}{100} \times \tfrac{5}{1000} = \tfrac{2135}{100000} = .02135$$

This method is not the one used by most text-book writers and teachers, but it is of value. If a teacher believes that decimal fractions are a transition from common fractions, and if he wishes his method to be in harmony with the manner in which the race developed the subject,—this is not always desirable,—he should use the above method.

The second method is based upon the theory that decimal

fractions originated from an extension of the number scale to the right of the units order. Before taking up the explanation by this method the effect of moving the decimal point a given number of places to the right or the left should be thoroughly understood. Ask the pupil to compare the values of the following numbers: 100, 10.0, 1.00, 0.1, and to state the effect of moving the decimal point to the left. Apply the principle to the following and see if it is true: 428, 42.8, 4.28, .428, .0428. Similarly, teach the effect of moving the decimal point to the right. The pupil should also be led to discover, or review, the principle that when one of two factors is multiplied by any number whatsoever and the other is divided by this *same* number, the product of the two factors is not changed. The truth of this principle may be made clear by illustrations.

Illustrations:

$$7 \times 4 = \tfrac{7}{10} \times 4 \times 10 = 28$$
$$5 \times 9 = \tfrac{5}{17} \times 9 \times 17 = 45$$

In the first illustration above we introduced ten as a factor in the numerator and also in the denominator; hence we introduce $\tfrac{10}{10}$, or 1.

If the pupil is required to find the product of $4 \times .07$ he may use the same principle as when he is required to find the product of $4 \times \$7$ or 4×7 books. He knows that 4×7 hundredths is 28 hundredths, or he may write it as .28. If the example requires the product of $.4 \times .07$ he may use the principles referred to above. He may multiply the .4 by 10 and divide the .07 by 10 without changing the value of the product. Hence, $4 \times .07 = 4 \times .007 = .028$. Similarly, $.03 \times .005 = 3 \times .00005 = .00015$. (In this case one factor is multiplied by 100 and the other factor is divided by 100.) Similarly, $32.54 \times .005 = 3254 \times .00005 = .16270$.

After the pupil has worked several examples he should, under the direction of the teacher, formulate his own rule for use, and he should see that whatever operation is necessary to make the multiplier an integer should be performed upon it, and the inverse operation should be performed upon the multiplicand. If the multiplier is multiplied by 10, 100, or 1000, the multiplicand should be divided by 10, 100, or 1000, respectively, in order that the value of the product may be unchanged.

Multiplication of decimals may be explained also in the following manner. Required to multiply .043 by .05. If we multiply .043 by 5, the result is .215, but this example differs from the given example only in that the multiplier is 100 times as large as the given multiplier (since $5 = 100 \times .05$). Therefore the product is 100 times too large, and the correct product is $\frac{1}{100}$ of .215, or .00215.

Whatever method is used should appear to the pupil as rational, and after a sufficient number of examples have been worked he should formulate the rule that in order to multiply two decimal fractions we proceed as in multiplication of integers and then point off in the product as many decimal places as there are in both the multiplicand and the multiplier. If the product does not contain as many figures as there are in the multiplicand and multiplier together, the necessary zeros must be prefixed to the product.

After the rule for multiplication has been rationally developed, the pupil should use it instead of developing the principle each time he wishes to find the product of two decimal fractions. The first essential is to make the process seem reasonable to the pupil, and the second is to acquire facility in the application of the process.

Division of decimals may be explained by first reducing the decimal fractions to common fractions and, after performing the division, expressing the quotient by decimal notation.

Illustrations:

$$.009 \div .03 = \tfrac{9}{1000} \div \tfrac{3}{100} = \tfrac{3}{10} = .3$$
$$.0045 \div .00005 = \tfrac{45}{10000} \div \tfrac{5}{100000} = 90$$

After working several examples, the rule which expresses the relation between the number of decimal places in dividend, divisor, and quotient should be stated.

Most authors and teachers explain division of decimal fractions by use of the principle that multiplying or dividing both dividend and divisor by the same number does not change the value of the quotient. This principle is readily understood from a few illustrations. For example, the quotient of $30 \div 6$ is not changed if both the 30 and the 6 are multiplied or divided by the same number.

$$30 \div 6 = (7 \times 30) \div (7 \times 6) \text{ or } 30 \div 6 = (\tfrac{1}{2} \text{ of } 30) \div (\tfrac{1}{2} \text{ of } 6)$$

If the pupil is required to work the example $.06 \div 3$, he knows the quotient just as he knows the quotient if asked to divide \$6 or 6 books by 3. If required to divide .06 by .3 he may use the principle above. Thus $.06 \div .3 = .6 \div 3 = .2$. (Both dividend and divisor were multiplied by 10.) Similarly, $.00015 \div .005 = .15 \div 5 = .03$. (Both dividend and divisor were multiplied by 1000.)

The pupil should be led to see that we first multiply both dividend and divisor by whatever will make the divisor an integer, and then perform the division.

After a sufficient number of examples have been worked by the above method, the pupil should formulate a rule for division of decimal fractions. Divide as in integers and point off in the quotient a number of decimal places equal to the excess of the number of decimal places in the dividend over the number of decimal places in the divisor.

This rule may also be deduced from the fact that division is the inverse of multiplication, and the dividend is the product of the divisor and quotient. Since the number of decimal places in the product equals the number in both multiplicand and multiplier, the number of decimal places in the quotient must equal the number in the dividend minus the number in the divisor.

In division of decimal fractions the quotient should be written above the dividend in such a manner that the decimal point of the quotient will be directly over the decimal point of the dividend.

Thus:

$$\begin{array}{r} 5.483 \\ \hline 8\overline{)43.864} \end{array}$$

Indicated multiplication and division containing decimals should generally be rewritten with the divisors changed to integers.

Thus

$$\frac{.47 \times 12.25 \times 1.03}{.23 \times 2.05 \times 6} = \frac{47 \times 1225 \times 1.03}{23 \times 205 \times 6}$$

(by multiplying both numerator and denominator by 100×100).

Contracted Methods

In certain work in science and in actual business activities operations are performed involving long decimal fractions, when the result is desired to only a few decimal places. Perfect measurements are rarely possible in science, and results beyond 3 decimal places are seldom required in business.

If a measurement is correct to only 3 decimal places,

any computation in which it is directly involved can not be correct to more than 3 decimal places. Much unnecessary labor is expended in some computations in seeking a product to more than 3 decimal places. In a commercial transaction such a result as $43.492736 would be considered as $43.49. It is always a waste of time to carry out results to a greater degree of accuracy than the data upon which the results are based or the situation demands.

The processes of multiplication and division involving decimal fractions may frequently be abridged by eliminating all unnecessary work. There is a real advantage in multiplication in beginning with the highest instead of the lowest order of units of the multiplier.

Suppose the product of 23.4271 by 1.3723 is required correct to 3 decimal places. In order to be sure that the result is correct to .001 it is wise to carry the partial products to .0001.

Full Form	Contracted Method
23.4271	23.4271
1.3723	1.3723
23.4271	23.4271
7.02813	7.0281
1.639897	1.6399
.0468542	.0469
.00702813	.0070
32.14900933	32.1490

Each partial product is increased by as many units as would have been carried from the rejected part: 0.5 is usually regarded as *one* to be carried.

Approximations are often desirable in division. The following illustration will indicate the method:[1]

---

[1] Beman and Smith, ''Higher Arithmetic,'' p. 11.

(1)   31416)329201(10.48
          3142   = approximately 10 × 3141 (6)
          ‾‾‾‾
          150
          126   approximately 0.4 × 314 (16)
          ‾‾‾
           24
           24   approximately 0.08 × 31 (416)

In contracted multiplication and division, considerable practice is necessary to give one confidence and to decide correctly what number should be carried. Contracted methods are not as generally used in the schools as their importance would justify. The European countries use them quite extensively. It is not desirable to emphasize these contracted methods in all communities, but in certain industrial centers some attention should be given to them.

Reduction of Decimal to Common Fractions

Decimal fractions are easily reduced to equivalent common fractions. To reduce a decimal fraction to a common fraction, omit the decimal point, write the denominator of the decimal, and then reduce the common fraction to its lowest terms.

Thus $.08 = \frac{8}{100} = \frac{2}{25}$.

This exercise will, in the case of some decimals, afford a review of complex fractions. Thus

$$.03\tfrac{1}{3} = \frac{3\frac{1}{3}}{100} = \frac{\frac{10}{3}}{100} = \frac{10}{300} = \frac{1}{30}$$

$$.02 7\tfrac{1}{2} = \frac{\frac{55}{2}}{1000} = \frac{55}{2000} = \frac{11}{400}$$

Reduction of Common to Decimal Fractions

The reduction of a common fraction to a decimal fraction may be explained in two ways.

First Method. A common fraction may be regarded as an indicated division. $\frac{2}{5}$ may be regarded as $2 \div 5$. Since we cannot take $\frac{1}{5}$ of 2 units exactly, we reduce the 2 units to the next lower denomination, thus making 20 tenths (2.0). $\frac{1}{5}$ of 20 tenths is 4 tenths (.4). Therefore $\frac{2}{5} = .4$.

Reduce $\frac{2}{7}$ to a decimal fraction. *Analysis.* $\frac{2}{7}$ of 1 unit equals $\frac{1}{7}$ of 2 units. We cannot take $\frac{1}{7}$ of 2 units, so we reduce the 2 units to lower denomination and get 20 tenths, or 200 hundredths, or 2,000 thousandths, etc. $\frac{1}{7}$ of 2,000 thousandths equals $285\frac{5}{7}$ thousandths, or $.285\frac{5}{7}$.

This process may be briefly performed by placing a decimal point after the numerator and dividing by the denominator.

Second Method. Perform any operation upon the fraction that will make the denominator a power of 10 but will not change the value of the fraction, then omit the denominator and express it by the position of the decimal point. Thus

$$\frac{3}{8} = \frac{3 \times 125}{8 \times 125} = \frac{375}{1000} = .375.$$

$$\frac{4}{125} = \frac{4 \times 8}{125 \times 8} = \frac{32}{1000} = .032.$$

(1)

$$\frac{7}{16} = \frac{7 \times 625}{16 \times 625} = \frac{4375}{10000} = .4375.$$

The first of these methods is easier for the pupil in most cases. The second method is often briefer but it requires more skill.

A common fraction cannot always be reduced to a pure decimal. Thus $\frac{1}{3} = 0.3\frac{1}{3} = 0.33\frac{1}{3} = 0.333\frac{1}{3}$. No common fraction can be reduced to a pure decimal if the denominator, when the fraction is reduced to its lowest terms, contains other prime factors than 2 and 5. This is true since ten is the product of 2 and 5 and any power of 10 is the product of an equal number of 2s and 5s. In reducing such common fractions as $\frac{1}{3}$ or $\frac{1}{7}$ to decimal fractions the division may be carried as far as is necessary and then the common fraction or the plus sign may be written. Thus $\frac{1}{7} =$ .14285$\frac{5}{7}$, or .14285 +.

Circulating Decimals

The reduction of certain common fractions to decimal fractions lead to the discovery of circulating decimals. Circulates are an outgrowth of the Arabic system of notation. The subject is very interesting, but it is more properly treated under infinite series in algebra than in arithmetic, and its discussion is omitted here for that reason.[1]

Aliquot Parts

The pupil should know the aliquot parts of 10, 100, and 1000 which are frequently used, and should learn their decimal equivalents. The equivalents for $\frac{1}{2}$, $\frac{1}{3}$, $\frac{1}{4}$, $\frac{1}{5}$, $\frac{1}{6}$, $\frac{1}{7}$, $\frac{1}{8}$, $\frac{1}{9}$, $\frac{1}{10}$, $\frac{1}{11}$, $\frac{1}{12}$, and for $\frac{2}{3}$, $\frac{3}{8}$, $\frac{5}{8}$, $\frac{7}{8}$, $\frac{5}{6}$ should be thoroughly mastered.

Sequence of Common and Decimal Fractions

Decimal fractions are taught in most schools of the United States in the fifth or sixth grades. Especial emphasis is usually put upon the subject in the 6th grade and percentage furnishes an opportunity to review decimals in the seventh or eighth grades. The sequence of common

---

[1] Teachers who are interested in the subject of circulating decimals will find it treated in the following references: Brooks, ''Philosophy of Arithmetic,'' pp. 460-485; McLellan and Dewey, ''The Psychology of Number,'' pp. 267-271; Showalter, ''Arithmetical Solution Book,'' pp. 49-55; Wentworth, ''Practical Arithmetic,'' pp. 354-356.

and decimal fractions is not yet settled. The general practice in the United States is to teach the simpler common fractions, such as ½, ⅓, and ¼ in the very early grades and to treat the entire subject of common fractions more or less formally in the fifth grade. Decimal fractions are rarely taught except as they are involved in the writing of United States money, until the fifth or sixth grade. No one seriously advocates the complete development of common fractions before any consideration is given to the subject of decimal fractions. If decimal fractions are regarded as a different notation for certain common fractions, the subject may be taught simultaneously with common fractions. The argument usually advanced by those who advocate the teaching of decimal fractions before common fractions is that the decimal fraction logically comes first since it is a natural outgrowth of our Arabic system of notation. Those who believe that common fractions should be taught first maintain that this sequence is in harmony with the historical development of the subjects and that the concept of the common fractions is simpler than that of the decimal fractions. It is further urged that pupils do not grasp the importance of decimal fractions and the simplifications which they effect until they have worked with common fractions with unlike denominators.

Problems Involving Decimals

All of the problems of decimal fractions, as of all other subjects, should be as practical as possible. They should have a direct and intimate relation to the daily activities in which decimal fractions are so extensively used and which come within the experience of the pupil.

Problems involving money, distance, speed and time; rainfall, temperature, crop yields, etc., are often of interest to pupils and furnish a good opportunity for computation in decimal fractions.

It has been asserted that the history of arithmetic has been a slow but well marked growth towards the decimal idea. The general adoption of decimal fractions has had a very marked effect upon several other topics of arithmetic. Decimal fractions have lessened the importance of the subjects of greatest common divisor and least common multiple and have simplified many of the calculations of science.

The importance of decimal fractions is recognized more today than ever before and the subject is one of the most important in arithmetic.

CHAPTER XV

PERCENTAGE

Percentage a Language Lesson

The subject of percentage is almost wholly a language lesson. No new mathematical principles are involved in any of its applications. When the child knows one-fourth, he knows twenty-five per cent, all but the name. The problem is to teach the pupil to think familiar ideas in a new language. Pupils should not feel that a change in terminology involves new mathematical processes. Some textbooks on arithmetic treat the subject of percentage as if it involved new mathematical principles. In such texts the subject is treated under various cases, rules and formulas. The best teachers of arithmetic to-day do not present the subject in this way. McClellan and Dewey state that the teaching of percentage by cases, rules and formulas is a mistake from both the theoretical and the practical points of view. It is a mistake on the theoretical side, "because it asserts or assumes a new phase in the development of number; on the practical side, because it substitutes a system of mechanical rules for the intelligent application of a few simple principles with which the student is perfectly familiar."[1] In actual business practice problems are not classified under their rules and cases. The terms base, rate, and percentage may be made to serve somewhat the same purpose in this subject that the terms

[1] McClellan and Dewey, "The Psychology of Number," p. 279.

numerator and denominator serve in the study of common fractions. The word rate may also properly be used when it refers to "rate of gain," "rate of interest," etc.

Relation between Percentage and Fractions

Percentage is a continuation of fractions. It is a special case of the subject and it may be made to afford excellent practice in enlarging the ideas of fractions and in securing greater facility in using them. The following illustration will serve to show the language change in the transition from fractions to percentage. A man had 700 chickens and sold one out of every two, or one out of every ten, or one out of every 50, or one out of every 100, or one out of every 350, how many were sold? The above problem would be classified under the subject of fractions. Since one out of two was sold, the number sold was $\frac{1}{2}$ of 700; or since 1 out of 10 was sold the number sold was $\frac{1}{10}$ of 700; or since 1 out of 100 was sold the number sold was $\frac{1}{100}$ of 700.

If we agree that the word *"per"* shall mean *"out of"* when used in this connection, the above statements would be: 1 per 2, or 1 per 10, or 1 per 100. If we substitute the Latin word "decem" for ten our statement becomes 1 per decem; if we substitute the Latin word "centum" for hundred, our statement becomes 1 per centum. If we abbreviate the word centum by cutting off the last two letters we have 1 per cent. We frequently abbreviate words in this way in arithmetic; for example, we write *"int."* for *"interest"* and *"fract."* for *"fraction."* One per cent, therefore, means 1 out of every 100. If a problem involving 1 out of every 100 is properly classified as a problem in fractions, it is certain that a problem involving one per cent is also a problem in com-

mon fractions. Smith states, in his Rara Arithmetica,[1]
that in certain old manuscripts % used to be written
"per 100." It became $\frac{c}{o}$ about 1650, and this became
0/0, then %. Percentage came in as a separate topic
about the beginning of the 19th century.

The mere fact that a quantity is measured off into hundredths instead of into any other possible number of parts
appears to be no valid reason for considering percentage
as a new phase in the development of number. Is the process of computing by hundredths "to be broadly distinguished as a mental operation from a process of computing by eighths, or tenths, or twentieths, or fiftieths? If
the difference between fractions and percentage is not a
difference in logical or psychological processes, but chiefly
a difference in handling number symbols, is it worth while
to invest the subject with an air of mystery and to invent,
for the edification of the pupil, from six to nine 'cases'
with their corresponding rules and formulas?"[2]

Applications

Percentage admits of applications in many fields, and
the increasing use of decimal fractions is increasing the
number of problems to which percentage is applied. In the
texts of a few decades ago almost all of the problems in
percentage had a financial basis, but to-day the best texts
contain numerous problems involving percentage when
money considerations are in no way involved. Some of
the applications are difficult to teach because the transactions are foreign to the experience of most of the pupils.
One of the chief difficulties which pupils encounter in

[1] Smith's "Rara Arithmetica," p. 440.
   The symbol ‰, read "per mil" and meaning "by the thousandths," is somewhat used in Europe, but is rarely used in the
United States.
   [2] McClellan and Dewey, "The Psychology of Numbers," p. 280.

studying the applications of percentage is the fact that they do not get a clear comprehension of the terms used. Unless the pupil has a clear conception of the terms employed he cannot form accurate judgments and conclusions. Too often the terms used are but "words, words, words," to the pupil. The various applications of percentage should be taught by correlating them as closely as possible with the actual situations in which they are used.

Anything that increases the interest of the pupil and helps him to associate the terms and the problems with actual situations in which they occur is worth while unless an excessive amount of time is expended. Any material that will vitalize the work may be used to advantage by the teacher. The pupil observes numerous business customs and procedures in his daily activities and his observations should be encouraged and made as accurate as possible.

It is important that the pupil should have an understanding of his natural environment and it is important that he should have an understanding of his business environment. An understanding of both natural and business environment is essential for the life work of a successful individual.

Whatever problems are used in percentage should be in harmony with actual business, industrial and scientific practice. All problems should be adjusted to present-day activities. Dewey's criticism that much that is learned in school cannot be applied in daily life is especially pertinent here. "The fact that the arithmetic of business is the arithmetic of common sense should not for a moment be lost sight of in drilling classes in this branch of our schools."[1] The teacher should seek to train the pupil to

[1] N. E. A. Committee Report on Business College Course, Vol. II., p. 2,163.

grasp conditions and to develop the power to apply the processes to actual business and industrial situations. All applications of percentage not consistent with modern practice should be omitted. One of the chief objects should be to develop commercial and industrial efficiency and social insight.

Equivalents of Certain Common Fractions

The teacher should impress upon the mind of the pupil the fact that "per cent" is identical in meaning with "hundredths." Six per cent of a quantity equals six one-hundredths of it. $10\% = \frac{10}{100} = .10$; $100\% = \frac{100}{100} = 1.00$; $357\% = \frac{357}{100} = 3.57$; $\frac{3}{8}\% = \frac{3}{8}/100 = .00375$.

Many problems involving the reduction of common fractions to per cent and of per cent to common fractions should be solved. Most of the work should be done orally and the pupil should be familiar with the fractional equivalents for the most common per cents of business. These should be memorized and drilled upon. The following are commonly used:

$6\frac{1}{4}\% = \frac{1}{16}$	$12\frac{1}{2}\% = \frac{1}{8}$	$25\% = \frac{1}{4}$	$62\frac{1}{2}\% = \frac{5}{8}$
$6\frac{2}{3}\% = \frac{1}{15}$	$14\frac{2}{7}\% = \frac{1}{7}$	$33\frac{1}{3}\% = \frac{1}{3}$	$66\frac{2}{3}\% = \frac{2}{3}$
$8\frac{1}{3}\% = \frac{1}{12}$	$16\frac{2}{3}\% = \frac{1}{6}$	$50\% = \frac{1}{2}$	$83\frac{1}{3}\% = \frac{5}{6}$
$10\% = \frac{1}{10}$	$20\% = \frac{1}{5}$	$37\frac{1}{2}\% = \frac{3}{8}$	$100\% = 1$

The equivalents for $\frac{1}{9}$, $\frac{1}{11}$ and $\frac{1}{14}$ are not frequently used.

There should also be drill in expressing decimal fractions as per cent. For example, $.1 = 10\%$; $.13 = 13\%$; $.9 = 90\%$; $.004 = \frac{4}{10}\%$, or $.4\%$; $.0005 = \frac{5}{100}\%$, or 0.05%; $0.00625 = \frac{5}{8}\%$; $.00\frac{2}{3} = \frac{2}{3}\%$; $5.24 = 524\%$.

The pupil should see that the reduction of a decimal fraction to per cent involves simply the reduction of the decimal fraction to hundredths and then instead of reading it *hundredths* we call it "per cent," which means hundredths.

The Three Problems of Percentage

In percentage and its application there are but three fundamental mathematical ideas involved and all of these are familiar to the pupil before he begins the formal study of the subject. The most complex problems in percentage contain no new mathematical principles. A pupil who has mastered the four fundamental operations with integers, and with common and decimal fractions, and, who can reduce common fractions to decimal fractions and vice versa, should have little or no difficulty with percentage. He may be unable to solve the problems, but his failure will be due to a failure to understand the terms employed and to think old ideas in terms of new symbols. Not the mathematics, but the terminology of percentage is the crux of the subject. Since only three fundamental mathematical ideas are involved in the problems of percentage it is economical of time and effort for the teacher to emphasize these three ideas before proceeding far with the subject.

The first fundamental idea is: To find any per cent of a given quantity. Find 7% of 45. The pupil knows that per cent means "hundredths" and from his study of decimal fractions he knows how to find .07 of a given quantity.

$$7 \% \text{ of } 45 = .07 \times 45 = 3.15$$
$$32 \% \text{ of } 372 = .32 \times 372 = 119.04$$

To find any per cent of a quantity, multiply the given quantity by the given per cent expressed as hundredths. *It is evident that a per cent of a given quantity may be regarded as the product of two factors.*

When per cents greater than 100% are involved the meaning "out of 100" is not so clearly understood. The pupil will easily understand the meaning of 200%, however, if the fact that % means hundredths is thoroughly understood.

Find $\frac{3}{8}$% of 416. $\frac{3}{8}$% = .00375; therefore $\frac{3}{8}$% of 416 = .00375 × 416 = 1.56. Or we may say, $\frac{3}{8}$% = $\dfrac{\frac{3}{8}}{100}$ = $\frac{3}{800}$; therefore $\frac{3}{8}$% of 416 = $\frac{3}{800}$ of 416 = $\frac{3}{100}$ of 52 = $\frac{156}{100}$ = 1.56.

It is evident that the application of this principle involves no new mathematical knowledge. The symbols involved are new, but the principle is not new.

The Second Fundamental Idea Is:

Find what per cent one number is of another.

Four is what per cent of eight?

In the preceding examples we were required to find a given per cent of a number. In this problem we are told that if we take a certain per cent of 8 the result is 4.

In other words, we have given the product of two factors, also one of the factors to find the other factor. This involves no new principle. If we divide the product, (4) by the given factor, (8), the result is the other factor. Since the problem asks for the result in terms of per cent (hundredths) we express the result not as $\frac{1}{2}$, but as $\frac{50}{100}$ or 50%.

Similarly:

7 is what % of 17?

? × 17 = 7. The unknown factor is found to be $\frac{7}{17}$, or .41$\frac{3}{17}$ or 41$\frac{3}{17}$%.

The Third Fundamental Idea Is:

Find a number when a certain per cent of it is known.

40% of a number = 10. Find the number.

Ten is the product of two factors, one of which, .40, is known. To find the other factor, divide the product, (10) by the known factor, .40, the result (25) is, therefore, the other factor, or the required number.

$$.40 \times ? = 10$$

Some teachers prefer to explain these processes by the use of unitary analysis. The above explanations, however, are brief and are easily understood by the pupils, furthermore they involve no new mathematical ideas and are serviceable.

The Solution of Problems

The pupil who has a thorough mastery of the preceding principles does not need to solve the problems of percentage by rules and cases. Sometimes he should use the purely fractional form, at other times the percentage form and at still other times he will combine the fractional and the percentage forms. "Practice in the application of the principles should enable the pupil to use all forms with equal facility and to determine in any given problem which of the forms will lead to the most elegant and concise solution." No set form should be required after the reasons for the various processes are understood, but the teacher should require the pupil to express himself in concise and accurate language. Many good teachers in the upper grammar grades require the solution of problems in percentage in step form with the actual work of the elementary operations omitted. Too great insistence upon minute details in written work may disgust and discourage some pupils. The important thing is to emphasize the relation of what is given to what is required.

If the problem states that 25% of a number is 18, the

pupil should immediately see that the number is 4×18, or 72. If the problem requires $33\frac{1}{3}\%$ of 480 the pupil should take $\frac{1}{3}$ of 480. Some object to the use of common fractions in problems which involve percentage. It must be remembered that percentage is studied in the grades because of its practical applications and the pupil should, whenever possible, solve the problems of percentage in the manner in which these problems are solved in actual life outside of the school. No business man would solve the following problem by keeping it in its percentage form. Twenty-five per cent of the cost of an article is $14, find the cost. The practical man would immediately find the cost by finding $4 \times \$14$. "The arithmetic of business is the arithmetic of common sense."

A Source of Failure in Percentage

One of the most frequent sources of error in the study of percentage is due to the failure of the pupil to keep clearly in mind *per cent of what*. In all early work in percentage teachers should require pupils to name the quantity upon which the per cent is based.

Questions such as the following will help to impress this important point.

1. What is 25% of 40?
2. 40 is 25% of what number?
3. 40 is what % of 25?
4. 40 is 25% greater than what number?
5. 25 is what % less than 40?
6. What number is 25% less than 40?
7. 40 is 25% less than what number?
8. 40 is what % greater than 25?

The following type of problem is especially good to impress the importance of keeping in mind the quantity that is to be made the basis in a problem involving **per-centage.**

Water expands 10% of its volume when it freezes. What per cent of its volume does ice contract when it melts?

The artificial problems found in many text-books are of the same general type. A's money is 25% more than B's. B's money is what per cent of A's?

Solution:

Since A's money is 25% of B's more than B's, therefore, A's money = 125% of B's money; therefore, **1% of** B's money = $\frac{1}{125}$ of A's money; therefore, 100% of B's money = $\frac{100}{125}$ of A's money, or $\frac{4}{5}$, or 80% of A's money.

The problem may also be solved by using common fractions throughout.

Types of Problems

Numerous problems which do not involve money should be given in percentage. The percentage of increase or decrease in population for various cities from 1900 to 1910; the percentage of games won or lost by the school teams; the percentage of attendance at school on a given day; the average heights of boys and of girls at different ages and the yearly percentage of increase; the average weight and strength at different ages; the per cent of pupils in each grade of the school based on the total school enrollment; the percentage of boys and girls in a given grade; the per cent of waste in cutting the largest possible circle out of a square; the per cent of decrease when various articles are dried or baked; the percentage of nutritive matter in various food articles; the per cent of yield of a given crop per acre as compared with other years; these suggest a few of the many problems not involving money that may be given to the pupils.

CHAPTER XVI

THE APPLICATIONS OF PERCENTAGE

PROFIT AND LOSS

The subject of profit and loss is one of the important applications of percentage. It is closely connected with business activities and the problems should be in harmony with business practice. The experiences of the pupils should be utilized as the basis for problems whenever this is possible, especially at the beginning of the work. The commercial and industrial interests of the community should be the source of numerous problems. In most of the problems involving profit and loss which the business man is called upon to solve the cost is known. Sometimes a merchant knows a possible selling price and must decide how much he can afford to pay to make a certain per cent. Some problems involving inverse cases may be properly introduced. It is said that many business men use a fixed selling price which is, for example, twenty-five per cent greater than the cost. When this is done the business man knows at once that his profits are 20% of his receipts, since a gain of 25% on the cost is 20% of the selling price. (25% of the cost is 20% of 125% of the cost).

No Rules or Cases Necessary

No formal rules or cases are necessary or desirable in solving the problems in profit and loss. The teacher should emphasize the fact that profit and loss are based upon the cost while commercial discount is based upon the list price. The problems in profit and loss may be grouped under five general headings.

I. Given the cost and the rate per cent of profit, or loss, to find the profit, or loss, and the selling price.

II. Given the cost and the profit, or loss, to find the rate per cent of profit, or loss, and the selling price.

III. Given the profit, or loss, and the rate per cent of profit or loss, to find the cost.

IV. Given the selling price and the rate per cent of profit or loss to find the cost.

V. Given the cost and the selling price to find the rate per cent of profit or loss.

Since profit and loss are always reckoned on the cost the procedure should be suggested by the nature of the problems.

Many problems should be solved without the use of paper or pencil. For example:

a. The cost is 8 cents and the selling price is 10 cents. Find % of gain.

b. The cost is 12 cents and the selling price is 12 cents. Find % of loss.

In problem "a" the pupil should see at once that the gain is 2 cents and the per cent gained is whatever per cent 2 cents (the gain) is of 8 cents (the cost). The only mathematics involved in such problems is a subtraction and "to find what per cent one quantity is of another quantity."

Type Solutions

The following solutions are intended to be merely suggestive. It is not assumed that they are the only correct ones. Teachers and pupils are cautioned against inaccurate expression in this subject.[1]

1. Cost = $420. Rate of gain = 20%. Find gain and selling price. 20% (or $\frac{1}{5}$) of $420 = $84 = gain.

Selling price = $420 + $84 = $504.

It is evident that the selling price might have been determined by finding 120% (or $\frac{6}{5}$) of $420.

2. The cost is $36. The loss is $4. Find selling price and per cent of loss.

The selling price is $36 - $4 = $32.

The per cent of loss is whatever per cent $4 (the loss) is of $36 (the cost).

The loss is $\frac{1}{9}$ or $11\frac{1}{9}$% of the cost.

3. The gain is $81. The rate of gain is $12\frac{1}{2}$%. Find cost and selling price.

$12\frac{1}{2}$% (or $\frac{1}{8}$) of cost = $81.

Therefore 100% (or $\frac{8}{8}$) of cost = $648.

Therefore selling price = $648 + $81 = $729.

4. The selling price was $560 and the rate of loss 20%. Find the loss and the cost.

Since the selling price was 20% of the cost less than the cost, it was 80% of the cost.

Therefore 80% of the cost = $560.

Therefore 100% of the cost = $700.

Therefore the loss = $140.

The above problem may also be solved as follows:

Let c = the cost

0.20c = the loss

0.80c = selling price = $560

therefore $\quad C = \dfrac{\$560}{.80} = \$700$

The introduction of the symbol "C" is a decided advantage in the solution of such problems and more emphasis should be given to such solutions in the schools.

---

[1] See chapter on Accuracy, p. 50-51.

5. The selling price was \$30; the cost was \$18. Find rate per cent of gain.

The gain was \$30 – \$18 = \$12.

The rate of gain is whatever per cent \$12 (the gain) is of \$18 (the cost). \$12 is $\frac{2}{3}$, or $66\frac{2}{3}\%$, of \$18. The gain is, therefore, $66\frac{2}{3}\%$.

In every problem in profit and loss in which the per cent of gain or loss is given, three per cents (or fractions) may always be found.

The first of these is always known. It is this, 100% of the cost = the cost.

The second of these is always given in the problem.

The third may always be found by adding or subtracting the first two.

For example, in the second problem above, it is known that 100% of cost = cost.

20% of cost = loss.

80% of cost = selling price.

Instead of using the per cent, the equivalent fractions may be used.

For example: The selling price is \$390; the rate of gain is 30%; find the cost.

$\frac{10}{10}$ of the cost = the cost.   (This is always known.)

$\frac{3}{10}$ of the cost = the gain.   (The problem states this.)

$\frac{13}{10}$ of the cost = selling price.   (By combining the first two.)

Therefore,

$\frac{13}{10}$ of the cost = \$390.

$\frac{10}{10}$ of the cost = $\frac{10}{13}$ of \$390 = \$300.

The above problem may be more briefly solved as follows:

$$c = \text{the cost}$$
$$1.30c = \$390$$
$$C = \frac{\$390}{1.30} = \$300.$$ The cost is \$300.

A man sold two lots for $150 each. He gained 25% on the first and lost 25% on the second. Find the entire gain or loss by the transaction.

Solution:

125% of cost of first lot = selling price of first lot. (Since it was sold at a gain of 25%.)

Therefore, 125% of cost of first lot = $150.

100% of cost of first lot = $120.

75% of cost of second lot = $150. (Since it was sold for $150 at a loss of 25%.)

100% of cost of second lot = $200.

Cost of both lots was $320. ($120 + $200.) Selling price of both was $300.

The loss was $20. The rate of loss was whatever per cent $20 (the loss) is of $320 (the cost). $20 is $6\frac{1}{4}$% of $320. The loss was, therefore, $6\frac{1}{4}$%. Such problems as this frequently confuse pupils because they do not keep in mind the *"per cent of what."*

I sold $\frac{2}{3}$ of an article for $\frac{3}{4}$ of the cost of the whole article. Find the rate per cent of gain or loss.

Since $\frac{2}{3}$ of the article sold for $\frac{3}{4}$ of the cost of the whole, $\frac{1}{3}$ of the article sold for $\frac{1}{2}$ of $\frac{3}{4}$, or $\frac{3}{8}$ of the cost of the whole; therefore $\frac{3}{3}$ of the article sold for $\frac{9}{8}$ of the cost of the whole.

The gain was, therefore, $\frac{1}{8}$ of the cost, or $12\frac{1}{2}$%.

The above solution may be considerably abridged by using "c" for cost and "s. p." for selling price.

Thus: $\frac{2}{3}$ s. p. = $\frac{3}{4}$c.

$\frac{1}{3}$ s. p. = $\frac{3}{8}$c.

$\frac{3}{3}$ s. p. = $\frac{9}{8}$c.

gain = $\frac{1}{8}$c. = $12\frac{1}{2}$%.

A careful study of the illustrative problems in profit and loss will corroborate the statement found on page 218 in regard to the mathematics involved in the applications of percentage.

The teacher should supplement the text-book problems in profit and loss by using many of the problems given under the subject of fractions.

COMMERCIAL DISCOUNT

The subject of commercial discount is usually taught in the seventh or eighth grade. It is intimately related to business activities. Trade or Commercial Discount is the deduction from the list price of an article. Since such discounts are usually made "to the trade" the term *trade discount* has arisen. Discounts are usually expressed in per cents or in fractions. The trend of business practice in recent years has been towards per cents that are easily reduced to common fractions, so that the discount can be computed easily. Thus a discount of $16\frac{2}{3}\%$ is easier to compute than one of 15%. Discounts are sometimes quoted in the following manner: 20 and $\frac{1}{3}$ off. This means that a discount of 20% is taken from the list price and then a discount of $33\frac{1}{3}\%$ ($\frac{1}{3}$) is taken from this remainder.

Reasons for Discounts

Among the most important reasons for granting discounts are: For cash; on account of fluctuations in the market price; to avoid the frequent publication of large catalogues; for large amounts purchased. Wholesalers and manufacturers usually publish a "list price" and then allow a certain per cent of discount "to the trade." Goods listed but not subject to discount are marked "net." A wholesale house usually has some such statements as the following upon its bills: Terms: "4 months, 30 days less

5%," or "30 days, 2% off 10 days," which mean that purchasers are entitled to a credit of 4 months, but will be allowed 5% discount if the bills are paid within 30 days; or that a credit of 30 days is allowed, but the purchaser will be allowed 2% discount if the bills are paid within 10 days. Sometimes goods are paid for before the shipping date for the goods has arrived. This is called "anticipating a bill." A discount equivalent to the current rate of interest is usually allowed on such a transaction. When possible, teachers should consult trade journals which quote prices and discounts.

Successive Discounts

When more than one discount is given, the first discount is reckoned upon the list price and the others are reckoned upon the successive remainders after the preceding discounts have been deducted. No two successive discounts are reckoned upon the same amount, hence successive discounts can never be added to find the equivalent single discount. Two successive discounts of 20 and 10 are not equivalent to a single discount of $20 + 10$. In all problems involving discounts the pupil must keep in mind the basis upon which the discount is to be computed.

Many teachers have the idea that successive discounts must be removed in the order in which they are quoted. The discounts may be removed in any order. The order does not affect the result. Successive discounts of 20, 10, and 5 will produce the same net price whether removed in the order just stated or in any of the following orders: 10, 5, 20, or 20, 5, 10, or 10, 20, 5, or 5, 10, 20, or 5, 20, 10. The truth of this statement may be easily verified.

Assume any marked price, for example, $360.

Suppose the discounts are quoted at 20, 10, and 5.

First Solution

$360 = list price or marked price
.20
$72.00, first discount
$360 – $72 = $288 = basis for second discount
.10
$28.80, second discount
$288 – $28.80 = $259.20 = basis for third discount
.05
$ 12.96, third discount
$259.20 – $12.96 = $246.24 = net price

Second Solution

$360 = list price or marked price
.10
$36.00, first discount
$360 – $36 = $324 = basis for second discount
.05
$16.20, second discount
$324 – $16.20 = $307.80, basis for third discount
.20
$ 61.56, third discount
$307.80 – $61.56 = $246.24 = net price

The net price is the same in each case.

It is apparent that the basis for the second discount might have been found by taking 80% of $360, instead of first finding 20% of $360 and subtracting this amount from $360. Similarly the basis for the third discount might have been found by taking 90% of the basis for the second discount and the net price might have been found by taking 95% of this amount. In other words, we first find 80% of the marked price, then 90% of this amount and finally 95% of this amount. The above operations may be indicated as follows:

.80 × .90 × .95 × \$360 = \$246.24. It is apparent that the result is not affected if the .90 and .80 are interchanged or any other interchange of the .80, .90, and .95 is made.

The fact that a change in the order of successive discounts does not change the list price may be very easily shown by use of algebraic symbols.

Finding Discount Equivalent to Several Single Discounts.

It is well for the teacher to know the short method for finding the single discount equivalent to two successive discounts.

Example: Find the single discount equivalent to the successive discounts of 20 and 10.

A discount of 20% leaves 80% of the list price to be paid. A further discount of 10% leaves 90% of the 80% of the list price, or it leaves 72% of the list price to be paid. Since 72% of the list price is to be paid, the two discounts of 20 and 10 must be equivalent to 28% of the list price (100% of list price − 72% of list price = 28% of list price.)

This equivalent may readily be found as follows: First add the two discounts; then multiply them and take .01 of their product; subtract this result from their sum; the remainder will be the single discount equivalent to the two successive discounts. Applying the rule just stated to the discounts, 20 and 10, we have the following:

20 + 10 = 30.   20 × 10 = 200.   .01 of 200 = 2.   30 − 2 = 28

A discount of 28 is equivalent to discounts of 20 and 10.

Find the single discount equivalent to the successive discounts of 15 and 5.

The sum is 20. The product is 75. .01 of the product

is .75. 20 – .75 = 19.25. Therefore, a discount of 19¼ is equivalent to discounts of 15 and 5.

If required to find the single discount equivalent to three successive discounts, first find the single discount equivalent to the first two discounts and then find the single discount equivalent to this result and the third discount.

For example: Find the single discount equivalent to the discount of 20, 10, and 5.

Discounts of 20 and 10 are equivalent to a single discount of 28. Discounts of 28 and 5 are equivalent to a single discount of 31.6. Therefore, discounts of 20, 10, and 5 are equivalent to a single discount of 31.6; similarly

Discounts of 20, 30 and 10 are equivalent to a single discount of 49.6.

Why Discounts are Quoted Separately

Pupils sometimes inquire why a given firm does not quote a single discount instead of quoting two or three successive discounts. Why shouldn't a firm quote a single discount of 31.6 instead of three discounts of 20, 10, and 5? The discounts are quoted separately in order that a purchaser may take advantage of one or two of the discounts even though he may be unable to take advantage of the three. The last discount quoted may be "for cash." The purchaser may be prepared to take advantage of the other discounts but he may be unable to pay cash for his goods. If the three discounts were grouped as a single equivalent discount a purchaser would not know the deductions to be allowed for specific reasons.

Illustrative Problems

The list price of some goods was $540. The discounts were 15, 10, and 6%. Find the net price.

First Solution

.85 ×.90 × .94 of $540 = $388.314 = net price.

Second Solution

Discounts of 15, 10, and 6 are equivalent to a single discount of 28.09. A discount of 28.09% leaves a net amount of 71.91% of the list price to be paid.

.7191 of $540 = $388.314 = net price

A few indirect problems in commercial discount are usually given in text-books. To illustrate:

What must be the marked price of goods to give discounts of 25%, 10%, and 10% and still realize $243?

$243 = net price

Discounts of 25, 10, and 10% are equivalent to a single discount of 39.25%

This leaves 60.75% of list price, which = net price..
Therefore, 60.75% of list price = $243.
Therefore, list price = $400.
The following is also a solution for this problem:

Let l = list price
Then .75 × .90 × 90 of l = $243
Therefore l = $\dfrac{\$243}{.75 \times .90 \times .75}$ = $400

A merchant buys goods at discount of 40% from the list price and sells at a discount of 30% of list price; what per cent does he gain on the cost?

The cost was 60% of the list price.
The selling price was 70% of the list price.
The gain was 10% of the list price.
Since % of gain is based on the cost, we must find what

% 10 per cent of the list price is of 60% of the list price (or the cost). It is ⅙ of it, or 16⅔% of it. Therefore, the gain is 16⅔% of cost.

At what % above cost must goods be marked in order to give a discount of 20% on the marked price and still make a profit of 30% on the cost?

Since a discount of 20% is given, the selling price is 80% of the marked price.

Since a profit of 30% is made, the selling price is 130% of the cost.

Therefore, 80% of marked price equals 130% of the cost.

1% of the marked price equals $\frac{13}{8}$% of the cost.

100% of the marked price equals $\frac{1300}{8}$% of the cost, or 162½% of the cost.

Therefore, the goods must be marked 62½% above cost.

It should be noticed that no mathematics is involved in the above problems except the four fundamental operations and the three principles of percentage previously referred to. The last problem solved involves the finding of a quantity (the list price) when a certain per cent of it is known.

Marking Goods

This is not a necessary part of a study of commercial discount or of profit and loss, but the pupils will be much interested in the subject and a brief consideration may be given to it.

Merchants frequently indicate the cost price and the selling price on each article. In order to conceal these from the customer, the merchant usually resorts to some symbols as a private mark. Usually some word or phrase containing ten different letters is selected and used as a "key." These letters are used to represent the nine digits and zero. Any word or phrase of ten letters or any ten arbitrary characters may be used as a "key" or "*cipher.*" An extra letter is used to prevent the repetition of a letter, and this is called a "*repeater.*" The repeater prevents giving any clew to the private mark, as it renders the deciphering more difficult. If the cost and selling price are both written on the same tag the selling price is usually written below and the cost above a horizontal line, but these positions are sometimes reversed. Sometimes the cost price is written in cipher known only to the proprietor.

Suppose the "key" is "Pay us often" and "x" is used as a repeater.

1	2	3	4	5	6	7	8	9	0	repeater
p	a	y	u	s	o	f	t	e	n	x

An article which cost $4.60 and is to sell at $5.40 would be marked

$$\frac{\text{u.on}}{\text{s.un}}$$

An article which cost $5.56 and is to sell at $6.50 would be marked

$$\frac{\text{s.xo}}{\text{o.sn}}$$

The following list of key words and phrases may be used by the class. Select a few of the "keys" and ask the pupils to express various costs and selling prices by using them.

Importance	Our Last Key
Charleston	Hard Moneys
Blacksmith	Buy for Cash
Republican	Cash Profit
Buckingham	The Big Four
Gambolines	United Cars
Authorizes	Black Horse

Bridgepost

Equinoctial

Frank Smith

Don't Be Lazy

Now Be Quick

Now Be Sharp

He Saw It Run

You Mark His

Market Sign

Big Factory

No Suit Case

Cumberland

COMMISSION

The subject of commission is studied in most schools in the seventh grade; in some it is taught in the sixth and in others in the eighth grade.

Necessity for Commission Business

The importance of the commission business in this country is largely the result of our industrial and commercial development. Certain regions are recognized as centers for particular products. Pittsburgh is the center of iron and steel industries; Chicago, Omaha, and Kansas City are live stock centers; Minneapolis is the center of enormous wheat and flour industries. The farmer who ships his wheat to Minneapolis or the local buyer who ships it for him cannot always take the shipment to the central market and sell it at the best price. Economic conditions demand that there shall be agents who shall represent the buyer and seller. The payment which this agent receives for his services is called a commission. The commission is usually a certain per cent of the amount of the sale or of the purchase; sometimes it is a specified sum for the performance of a certain service.

The commission agent transacts business for and in the name of another. The articles which he buys or sells may be products of the soil, such as wheat, corn, or cotton; or they may be live stock, farm or city property, food stuffs, or numerous other things. The compensation paid to insurance agents, book agents, auctioneers, buyers of farm stock and tax collectors is usually called a commission. A collector of accounts is also said to receive a commission, which is based on the amount collected. Commission merchants are responsible to their principals for the value of goods sold by them on credit. It is not the custom for the commission merchant to charge a separate rate for assuming this risk, but the rate of commission is large enough to cover it. This is one respect in which a commission merchant differs from an agent.

Local Basis for Commission

The subject of commission may be made interesting to both city and country pupils if the introductory problems are based on the sending of farm products to the city. In almost every locality some one is engaged in the commission business and these local cases should be utilized, when possible, in studying the subject. Most of the problems should involve direct operations only and all of them should be of the types that are in harmony with actual business practice. The tax collector of the community may be used as an illustration of one whose compensation is a commission. Pupils should find out the basis upon which his commission is computed. The technical terms of the subject should be understood by the pupil before he attempts to solve the problems.

Technical Terms of the Subject

Such terms as agent, principal, consignment, consignor, consignee, remittance, net proceeds, account sales and account purchased are technical words of the subject. These terms are defined in the text-books and need no comment here. The language of commission usually causes the pupil more difficulty than the mathematics of the subject.

The mathematics of commission involves only the fundamental problems of percentage.

Illustrative Problems

The problems which follow illustrate the various types that occur in the subject of commission.

A man places a claim of $1800 in the hands of an attorney for collection. The attorney succeeds in collecting only 60 cents on the dollar. Find the amount of the attorney's fee if he receives a commission of $2\frac{1}{2}\%$.

$$
\begin{array}{l}
\$1800 \text{ amount of claim} \\
\quad\ \ .60 \\
\hline
\$1080.00 \text{ amount collected} \\
\quad\ \ .025 \\
\hline
\quad 5400 \\
\quad 2160 \\
\hline
\$27.000 \text{ amount of fee}
\end{array}
$$

I paid an agent $55.80 for buying wheat on a commission of 3%. Find the amount spent for wheat.

3% of the purchased price = the commission.

$55.80 = the commission.

∴ 3% of the purchased price = $55.80.

100% of the purchased price = $1860.

The proceeds from the sale of a house were $3504. The real estate agent received a commission of 4%. What was the selling price of the house?

Since the commission was 4% of the selling price,

∴ the proceeds were 96% of the selling price,

96% of the selling price = $3504

100% of the selling price = $3650.

I paid my agent 5% for selling corn and 2% for investing the net proceeds in wheat. What was the selling price of the corn if his entire commission was $164.50?

100% of selling price of corn = selling price of corn.

95% of selling price of corn = net proceeds from sale. (This includes the commission paid for buying wheat.)

102% of price paid for wheat = 95% of selling price of corn.

1% of price paid for wheat = $\frac{95}{102}$%. of selling price of corn.

2% of price paid for wheat = $\frac{95}{51}$% of selling price of corn. (This was the second commission.)

The sum of the two commissions was 5% of selling price of corn + $\frac{95}{51}$% of selling price of corn, or 6$\frac{44}{51}$% of selling price of corn.

∴ 6$\frac{44}{51}$% of selling price of corn = $164.50.

∴ 100% of selling price of corn = $2397.

Find the amount of a sale if an agent charged a commission of 3%, $12.50 for drayage, $10.50 for storage, and $3.50 for insurance. The net proceeds of the sale were $1389.70.

The charges, exclusive of commission, were $26.50.

∴ the amount of sale less the commission was $1416.20.

∴ 97% of the amount of the sales = $1416.20.

100% of the amount of the sales = $1460.

My agent sold goods for $2260; if the commission was $90.40 what was the rate per cent of commission?

$90.40 is 4% of $2260, therefore, the commission was 4%.

SIMPLE INTEREST

Pupils sometimes fail to master the subject of interest, but the failure may usually be traced to the lack of an accurate understanding of the terms used and of an acquaintance with business procedure rather than to any mathematical difficulties involved. There is no problem in simple interest the solution of which requires a degree of mathematical knowledge not in the possession of the pupil who is prepared to begin a formal study of the subject. Simple interest is an easy application of percentage with time as an important factor.

Numerous definitions have been suggested for the term "interest." Some text-books define it as money paid or charges for the use of money. The statement that "interest is money rent" is probably as good as any that have been proposed.

The pupil should understand how men, when lending money, require a certain payment for the use of the money. The teacher should make clear to the pupils how a man can afford to borrow a given sum for a year, be security for the amount borrowed, and at the end of the year pay back to the lender not only the amount originally borrowed, but an additional amount, which is called interest. The sum upon which interest is based is called *principal,* to distinguish it from the interest, which is of subordinate importance.

History of Interest

A brief historical survey of the subject of "interest" is quite profitable. That the practice of receiving interest should ever have been regarded as immoral and as a wrong to society, seems strange to us because the custom is so well established; but the propriety of such a charge has been frequently questioned.

Interest, or usury, as it was formerly called, was charged in the time of the Babylonians. From numerous references in the Bible we conclude that among the early Hebrews it was unlawful to charge money for the use of money. In later years it was considered lawful to charge a stranger usury (or interest). Finally, it was regarded as lawful to accept usury (or interest) from anyone. "Thou oughtest therefore to have put my money to the exchangers, that at my coming I should have received mine own with usury.''[1] Interest rates in Greece varied from 12% to 18%; in Rome 48% was allowed in Cicero's time, and 6% in the time of Justinian. The medieval church was generally hostile to the practice of charging interest. Italy was at one time the great financial center of Europe, and the practice of charging interest was common there. In 1552 a law was passed in England prohibiting the charging of interest, and the practice was declared to be ''contrary to the will of God.''

Usury

The term ''usury'' means etymologically ''the use of a thing.'' The word was originally applied to the legitimate payment of money for the use of money, and was synonymous with the term ''interest'' as we use the term to-day.

In most of the countries where usury or interest was permitted, laws were passed which limited the rate that might be charged. The evasion of these laws by charging excessive usury led to the current use of the term. To-day the word usury means the rate or amount of interest in excess of that permitted by law. In the United States the maximum rates of interest are usually specified by the states. The maximum contract rate in New York, Pennsylvania, and New Jersey is 6%; in Michigan and Illinois it is 7%; in Ohio and Indiana it is 8%. If a citizen making a loan in Illinois, where the maximum rate is 7%, attempts by legal procedure to collect more than 7%, he is guilty of usury. In Ohio any rate higher than 8% would be legally considered as usury. The penalty for charging usury differs in various states. In some states the offender loses all of the interest; in others he loses

[1] Matthew 25:27.

both principal and interest; some states provide no penalty for usury. In several states the usury laws have been repealed and the tendency is to allow an open market for capital. Teachers may borrow a statute from a justice of the peace and read the law on interest and usury to the pupils.

The following table shows the interest rates and penalties for usury prevailing in the states and territories:

States and Territories.	Legal Rate. —Per	Maximum Rate Allowed. cent—	Penalty for Usury.
Alabama	8	8	Forfeiture of all interest
Alaska	8	12	
Arizona	6	Any	None
Arkansas	6	10	Forfeiture of principal and interest
California	7	Any	None
Colorado	8	Any	None
Connecticut	6	6	None
Delaware	6	6	Forfeiture of double amount of loan
Dist. of Columbia	6	10	Forfeiture of all interest
Florida	8	10	Forfeiture of all interest
Georgia	7	8	Forfeiture of all interest
Hawaii	6	12	
Idaho	7	12	Forfeiture of three times the excess of interest over 12%
Illinois	5	7	Forfeiture of all interest
Indian Territory	6	8	
Indiana	6	8	Forfeiture of excess of interest over 6%
Iowa	6	8	Forfeiture of all interest and costs
Kansas	6	10	Forfeiture of double the excess of interest over 10%
Kentucky	6	6	Forfeiture of excess of interest
Louisiana	5	8	Forfeiture of all interest
Maine	6	Any	None
Maryland	6	6	Forfeiture of excess of interest
Massachusetts	6	Any	None

States and Territories.	Maximum Legal Rate. Rate. Allowed. —Per cent—		Penalty for Usury.
Michigan	5	7	Forfeiture of all interest
Minnesota	6	10	Forfeiture of contract
Mississippi	6	10	Forfeiture of interest
Missouri	6	8	Forfeiture of all interest
Montana	8	Any	None
Nebraska	7	10	Forfeiture of all interest and cost
Nevada	7	Any	None
New Hampshire	6	6	Forfeiture of three times the excess of interest
New Jersey	6	6	Forfeiture of all interest and costs
New Mexico	6	12	None
New York	6	6	Forfeiture of principal and interest
North Carolina	6	6	Forfeiture of double the amount of interest
North Dakota	7	12	Forfeiture of all interest
Ohio	6	8	Forfeiture of excess over 8%
Oklahoma	7	12	
Oregon	6	10	Forfeiture of interest, principal and costs
Pennsylvania	6	6	Forfeiture of excess of interest
Philippine Islands	6	Any	
Porto Rico	12	12	
Rhode Island	6	Any	None
South Carolina	7	8	Forfeiture of all interest
South Dakota	7	12	Forfeiture of all interest
Tennessee	6	6	Forfeiture of excess of interest
Texas	6	10	Forfeiture of all interest
Utah	8	Any	None
Vermont	6	6	Forfeiture of excess of interest
Virginia	6	6	Forfeiture of excess of interest over 8%
Washington	10	12	Forfeiture of double illegal interest
West Virginia	6	6	Forfeiture of excess of interest
Wisconsin	6	10	Forfeiture of all interest
Wyoming	8	12	None

Why Maximum Interest Laws Were Enacted

Teachers should consider the conditions which gave rise to the enactment of maximum interest laws by many of the states. The laws allow an open market to capital which is invested in houses or farms or merchandise. An owner is permitted to secure for his property whatever rent the laws of supply and demand will permit. If a man has his capital in the form of money and wishes to rent it, that is, to charge interest for it, why should the state specify the maximum rent which he may legally receive? Is the borrower any more at the mercy of the unprincipled money lender than the renter is at the mercy of those who seek to charge exorbitant rents? Is he not liable to be a victim of extortion in either case? What reason is there for regulating interest rates that will not apply to the regulating of rents for houses or farms? The answer to this question will doubtless be suggested to the thoughtful teacher when he considers those who needed protection from exorbitant interest rates when the laws were enacted and those who are the great borrowers of capital to-day.

Many states not only specify the maximum rate of interest that may be legally collected, but the "legal rate" is also determined by legislative enactment. The "legal rate" means that rate of interest which may be legally collected when the words "with interest" are included in a note but no rate of interest is specified. The legal rate is also the rate which may be collected on a note without interest which is not paid when due. Such a note draws the legal rate of interest from the date of maturity until it is paid. It will be seen from the preceding table that in some states the legal rate is the same as the maximum rate; in others it is less.

How Interest Rates Are Determined

The pupils should consider the various factors which determine interest rates. The relation of interest rates to the law of supply and demand for money should be pointed out. Other things being equal, interest rates are low when the amount of money to be lent exceeds the demands of those who wish to borrow; rates are high when the demands of the borrowers exceed the amount of money to be lent. Especial attention should be directed to the security of the loan as a factor in determining interest rates. The United States Government can borrow large sums of money at low rates of interest. An unstable government must pay high interest rates. It is the duty of the teacher to impress upon the pupils the fact that very high rates of interest are often synonymous with poor security. The first thing to be investigated in making a loan is not the rate of interest that is charged, but the security of the loan. It would not be correct to say that high rates of interest are always directly associated with poor security, for where profits upon capital are large, the rates of interest are high as a result of the law of supply and demand. However, the teacher should caution the pupil to investigate with more than usual care the security of any loan when very high returns are promised. In this connection some consideration should be given to the pernicious practice of the "loan sharks." These people often secure the payment of exorbitant rates of interest and the unfortunate man or woman who becomes financially obligated to a "loan shark" finds it next to impossible to discharge his so-called obligation and free himself from the clutches of his oppressor. In recent years the courts are becoming more active in condemning the impossible contracts which these lenders induce their victims to sign.

A third factor that determines interest rates is the time
for which the loan is made. The rate is usually lower
upon a loan for a long period than for a short period. The
amount of the loan is also a factor in determining interest
rates. The rate for a small loan is often higher than for a
large one.

Days of Grace

Owing to the rapid growth of banking facilities, short
term notes have become so general that the teaching of
interest is gradually becoming simplified in the schools. It
is now customary to pay interest every 30, 60, or 90 days,
or else every year. Days of grace have been abolished in
many of the states because the development of banking
facilities has rendered them unnecessary. The teacher
should ascertain whether days of grace are still permitted
in his state; and if not permitted, they should be omitted
from all problems, irrespective of the statement in the
text-book.

Kinds of Interest

Since the year contains approximately $365\frac{1}{4}$ days, there
is, strictly speaking, no such thing as exact interest. However, the name exact interest is applied to simple interest
computed upon the exact number of days of each month
and considering the year as consisting of 365 days.

In computing bankers' interest the exact number of days
of each month is reckoned, but the 360-day year is used.
In common interest 360 days are considered a year, 30
days a month, and a month $\frac{1}{12}$ of a year.

Exact interest is used by the United States Government,
by some large banks, and in finding the interest on foreign
money. Bankers' interest is used by most banks and by
business men in finding the interest on short time notes.

Common interest is used in finding the interest on notes

and debts that bear interest for longer periods of time—usually when the time is a year or more.

Exact interest for any number of days may be found by taking $\frac{72}{73}$ of the amount of the common interest. ($\frac{360}{365}$ is $\frac{72}{73}$ of $\frac{365}{365}$). Common interest may be found from exact interest by increasing the exact interest by $\frac{1}{72}$ of itself. ($\frac{365}{365}$ is $\frac{73}{72}$ of $\frac{360}{360}$).

If we compare the exact, bankers', and common interest on a given principal at a given rate for an interval of several months, we will find that the bankers' interest is slightly higher and the exact interest is slightly lower than the common interest.

Methods of Computing Interest

There are numerous methods for computing simple interest. One or more of the following methods are usually considered in text-books on arithmetic: Six Per Cent, Aliquot Part, Twelve Per Cent, Formula, Cancellation, One Per Cent, Thirty-six Per Cent, Dollar, Month, One Day, Bankers', Sixty Day. Pupils should be taught only one method besides the general method for computing interest. The attempt to develop a mastery of several methods usually results in confusion. It is much better to secure a mastery of one good method than to have a partial mastery of several. There is some difference of opinion as to the best single method for computing interest, but the tendency in recent years seems to be towards the six per cent, the aliquot part, and the formula methods, although a number of the others have commendable features.

The Six Per Cent Method

The six per cent method is one of the shortest for finding the interest on $1 for a given number of days. Since it is based on a year of 360 days, it is somewhat inexact. However, in many financial transactions the difference between the interest computed on the basis of a year of 360 days and that computed on the basis of 365 days is negligible.

The six per cent method is based on the following facts:

Interest on $1 at 6% for 1 year (360 days)=$.06.

Interest on $1 at 6% for 2 mo. ($\frac{1}{6}$ of 1 yr.)=$.01.

Interest on $1 at 6% for 1 mo. ($\frac{1}{2}$ of 2 mo.)=$.005.

Interest on $1 at 6% for 6 da. ($\frac{1}{5}$ of 1 mo.)=$.001. ·

Interest on $1 at 6% for 1 da. ($\frac{1}{6}$ of 6 da.)=$.000$\frac{1}{6}$.

It is readily seen that the interest on $1 at 6% for 1 month is $\frac{1}{2}$ cent. The interest on $1 at 6% for any number of months is half as many cents as there are months. Thus the interest for 8 mo. is $.04; for 7 mo. it is $.035; for 11 mo. it is $.055.

It is evident also that since the interest on $1 at 6% for 1 day is $\frac{1}{6}$ of a mill, the interest on $1 at 6% for any number of days will be $\frac{1}{6}$ as many mills as there are days. Thus the interest on $1 at 6% for 18 da. is $.003; for 21 da. it is $.0035.

The above facts enable us to determine readily the interest on $1 at 6% for any number of months or days. Required to find the simple interest on $480 for 7 mo. 18 da. at 6%.

The interest on $1 at 6% for 7 mo. 18 da. is $.038.

∴ the interest on $480 at 6% for 7 mo. 18 da. is 480 × $.038 = $18.24.

If the rate above were 4% we would take $\frac{4}{6}$ of the interest just found. If the rate were 3$\frac{1}{2}$% we would divide the above interest by 6 (which would give the interest at 1%) and then multiply this result by 3$\frac{1}{2}$.

The Aliquot Part Method

In computing interest by the Aliquot Part Method the time is separated into two or more parts so that each part less than a year is a unit fraction of some preceding part. One illustration will make the method clear.

Required to find the simple interest on $480 for 1 yr. 7 mo. 18 da. at 5%.

$480
.05
$$\overline{\$24.00}\ \text{Int. for 1 yr.}$$

6 mo. $= \frac{1}{2}$ yr.	$12. Int. for 6 mo.
1 mo. $= \frac{1}{6}$ of 6 mo.	$ 2. Int. for 1 mo.
15 da. $= \frac{1}{2}$ mo.	$ 1. Int. for 15 da.
3 da $= \frac{1}{10}$ mo.	$ 0.20 Int. for 3 da.

$$\overline{\$39.20}\ \text{Int. for 1 yr. 7 mo. 18 da.}$$

The Aliquot Part Method is not applicable when accurate interest is desired.

The Formula Method

The chief objection urged against the formula method is that the principles underlying the calculation of interest are so elementary that there is no occasion to substitute mechanical rules for the application of a few simple principles. If the pupil knows the general formula $I = \frac{p\ r\ t}{100}$ and can solve a simple equation he can use the formula to advantage. If the habit of employing equations in the solution of problems has been formed, such formulas as the above may be used to advantage.

Problems in Simple Interest

Many of the problems in simple interest should be solved orally and the pupil should be encouraged to use the shortest method to secure the desired result. In general, it is better to postpone the consideration of short methods until the pupils are reasonably familiar with the general methods.

The six per cent method may be considerably abridged, and there is no reason why the short form should not be taught to the pupil after he is familiar with the more expanded form.

If required to find the simple interest on $420 for 5 mo. 18 da. at 6%, we may proceed as follows:

SOLUTION:

$$\frac{168 \times \$420}{6 \times 1{,}000} = \$11.76$$

The 168 is derived from the reduction of 5 mo. 18 da. to days. We divide by 6 because the principle requires us to take $\frac{1}{6}$ of the number of days. We divide by 1,000 because, having taken $\frac{1}{6}$ of the number of days, we must use the result as mills, and mills stands in thousandths place.

What is the simple interest on $480 at 7% for 5 mo. 12 da.?

$$\frac{162 \times 7 \times \$480}{6 \times 6 \times 1000} = \$15.12$$

We divide by the first six because we must take $\frac{1}{6}$ of the 162 days; by the second 6 in order to find the interest at 1%; we multiply this result by 7 in order to obtain the interest at 7%.

One more problem will illustrate sufficiently this abridged form. Find the simple interest on $840.24 for 3 mo. 24 da. at 5%.

3 mo. 24 da. = 114 da.

$$\frac{114 \times 5 \times \$840.24}{6 \times 6 \times 1000} = \$13.303$$

One advantage of this method is the opportunity afforded for cancellation.

Computing Interval between Dates

In computing the time between two given dates, the teacher should require the pupils to use the method prevalent in the community. In some localities the time is computed in days when the interval is less than a year, and in years, months, and days when the time is greater than a year. In other localities the time for periods greater than a year is found in years and days. The time from April 14, 1910, to November 8, 1912, may be computed by any of the following methods:

$$\begin{array}{r} \text{(a)} \quad 1912 \quad 11 \quad 8 \\ \underline{1910 \quad 4 \quad 14} \\ 2 \quad 6 \quad 24 \end{array}$$

The result is 2 yr. 6 mo. 24 da.

(b) The time from April 14, 1910, to April 14, 1912, is 2 years. From April 14 to October 14 is 6 months. From October 14 to November 8 is 25 days. The result is 2 yr. 6 mo. 25 da.

(c) From April 14, 1910, to April 14, 1912, is 2 years. From April 14 to November 8 (counting the exact number of days in each month) is 208 da., or 6 mo. 28 da. The entire time is therefore 2 yr. 6 mo. 28 da. There are sections of the United States in which each of the three methods is recognized as correct. In some states both the day a note is given and the day it matures are considered in computing the time.

Some texts contain a table showing the number of days from any day of one month to the same day of any other month within a year. Such a table is found on page 275.

Interest Table

If time permits teachers may interest pupils in constructing a portion of an interest table, such as is frequently employed by bankers. A portion of such a table is given below.

2 months 4%

Total days.	$1000	$2000	$3000	$4000	$5000	$6000	$7000	$8000	$9000
60.....	6.66	13.33	19.99	26.65	33.33	39.99	46.66	52.22	59.99
61.....	6.77	13.55	20.33	27.10	33.89	40.66	47.43	54.21	60.99
62.....	6.88	13.77	20.66	27.55	34.44	41.32	48.21	55.10	61.99

The table shows that the interest on $3000 for 60 days at 4% is $19.99. The interest on $300 for 60 days is $1.999 and on $30 it is $.1999. The interest on $7000 for 62 days at 4% is $48.21. By moving the decimal point the interest can easily be found on $700, $70, $7, or $0.70 at 4%.

By using the table we may find the interest on $5486 for 61 days at 4% as follows:

The interest on $5000 for 61 days at 4% is $33.89.

The interest on $400 for 61 days at 4% is $2.71.

The interest on $80 for 61 days at 4% is $0.5421.

The interest on $6 for 61 days at 4% is $0.0406.

Therefore, the interest on $5486 for 61 days at 4% is $37.1827, or $37.18.

Similar tables may be constructed for any number of days and at any desired per cent. Such tables are much used by banks, insurance offices and trust companies. The tables are arranged in convenient form and they greatly lessen the labor of computing interest. Several different arrangements of interest tables are published.

Indirect Problems

The "indirect" problems in interest have a very limited application and many of them should be omitted. The important thing is to find the interest. There is a growing tendency to devote attention to the real problems only in studying interest and to use the time thus saved in writing promissory notes and in familiarizing the pupil with other business procedures. Genuine mercantile transactions supply problems that are sufficiently complex for the average pupil. It is certain that the indirect problems in interest, if taught, should not be taught by rules. It is practically useless for the pupils to work the problems in a mechanical way. The problems, if considered at all, should be taught so as to afford a training that is worth while. A few of the indirect problems will be considered.

1. The interest on $420 for 2 years, 3 months is $47.25, what is the rate?

a)

$$2 \text{ yr. } 3 \text{ mo.} = \tfrac{9}{4} \text{ yr.}$$
The interest for $\tfrac{9}{4}$ yr. = $47.25
The interest for 1 yr. = $\tfrac{4}{9}$ of $47.25 = $21
$\therefore$ r% of $420 = $21
$\therefore$ r% = $\dfrac{\$21}{\$420} = \tfrac{1}{20} = 5\%$

The above problem may be solved as follows:

b) The interest on $420 for 2 years, 3 months at 1% = $9.45; therefore, to produce $47.25 in the same time the rate must be as many times 1% as $47.25 is times $9.45. $47.25 is 5 times $9.45; hence the required rate must be 5%; or it may be solved as follows:

c)

$$\text{Let } r\% = \text{the rate}$$
$$r\% \times 2\frac{1}{4} \times \$420 = \$47.25$$
$$\therefore r\% = \frac{\$47.25}{\$9.45} = 5\%$$

2. How long will it take the interest on $480 at 6% to equal $48?

a)

The interest for 1 yr. = .06 of $480

The interest for t yr. = t × .06 of $480

$$\therefore t \times .06 \text{ of } \$480 = \$48 \quad \therefore t = \frac{\$48}{\$28.80} = 1\frac{2}{3}$$

∴ the time is $1\frac{2}{3}$ yr., or 1 yr. 8 mo.

The above problem may be solved as follows:

b) The interest on $480 for 1 year at 6% is $28.80. To produce $48 interest at the same rate, the time must be as many times one year as $48 is times $28.80. $48 is $1\frac{2}{3}$ times $28.80; hence, the time is $1\frac{2}{3}$ year, or 1 year and 8 months.

3. On what sum of money will the interest for 1 year, 4 months, at 6%, equal $57.60?

a) Let $p = the principal.

The interest on $p for $\frac{4}{3}$ yr. at 6% is $57.60

Therefore the interest on $p for 1 yr. at 6% is $\frac{3}{4}$ of $57.60, or $43.20

$$\text{therefore} \quad \$p \times .06 = \$ \ 43.20$$
$$p = \frac{\$43.20}{.06} = \$720.$$

or we may say:

(b)

$$\tfrac{4}{3} \times .06 \times \$p = \$57.60$$
$$\therefore \; p = \frac{\$57.60}{\tfrac{4}{3} \times .06} = \frac{\$57.60}{.08} = \$720$$

or as follows:

c) A principal of \$1 will produce \$.08 interest in 1 yr. 4 mo. at 6%. To produce \$57.60 interest the principal must be as many times \$1 as \$57.60 is times \$.08. \$57.60 is 720 times \$.08, hence the required principal is \$720.

4. Find the principal which will amount to \$562.68 in 8 mo. 12 da. at 6%.

a)

$$p = \text{principal}$$
$$p + .042p = \$562.68$$
$$\therefore \; 1.042p = \$562.68$$
$$p = \frac{\$562.68}{1.042} = \$540$$

(.042 is found by getting the interest on \$1 at 6% for 8 mo. 12 da.)

or as follows:

4. (b) A principal of \$1 will amount to \$1.042 in 8 mo. 12 da. at 6%. To amount to \$562.68 the principal must be as many times \$1 as \$562.68 is times \$1.042. \$562.68 is 540 times \$1.042; hence, the required principal is \$540.

5. In what time will any principal double itself at 5% simple interest?

To double itself a principal must gain 100% of itself. Since in 1 yr. the principal gains 5% of itself, it will require 20 years to gain 100% of itself.

6. At what rate will any principal double itself in 25 years?

Since a principal gains 100% of itself in 25 yr.; $\therefore$ it gains $\tfrac{1}{25}$ of 100% of itself in 1 yr. The yearly gain is 4%, which is, therefore, the rate.

ANNUAL INTEREST

Annual interest is interest payable annually or at any other regular interval, as semi-annually, or quarterly. If annual interest is not paid when due it draws simple interest from the time it becomes due until it is paid. Annual interest may be defined as simple interest on the principal, increased by the simple interest on each interval's interest from the close of the interval to the time of settlement. The interval may be one year or one-half year, or one-fourth of a year, or any other specified period. If the interest is payable semi-annually or quarterly it is computed in the same manner as when it is payable annually.

Since the intervals involved in this type of interest are sometimes fractional parts of a year the term "periodic interest" is frequently used instead of annual interest.

In many states unpaid interest does not draw interest until settlement, but annual interest is legalized in Michigan, Ohio, Wisconsin, Vermont, New Hampshire, and Iowa. In Pennsylvania, Georgia, Illinois, and Indiana it is legal by special contract only. It is the custom in some parts of the country to draw up a note for the principal without interest for the specified time and interest notes which mature at the time each interest payment is payable.

These interest notes provide for the payment of simple interest on all unpaid interest.

If annual interest is not permitted in a given state, it should not be taught in the schools. If it is permitted in the state, the teacher should ascertain the rate at which it is legal and should use that rate in the problems.

Illustrative Problems

The solution of one problem will suffice to illustrate this type of interest.

Indianapolis, Ind., June 18, 1913.

$600.

Four years after date I promise to pay to the order of F. S. Lint Six Hundred Dollars, value received, with interest at 5%, payable annually.

THOS. W. BRIGGS.

If no interest is paid until this note matures, how much is then due?

Solution. The interest on $600 for 4 years at 5% is $120.

The interest for the first year is $30 and this will draw interest at 5% for 3 years. (Since the note does not mature for 3 years after the first interest is due.) The interest for the second year will draw interest for 2 years and the interest of the third year for 1 year. Thirty dollars will, therefore, draw interest at 5% for 3 years plus 2 years plus 1 year, or 6 years. The interest on $30 for 6 years at 5% is $9. The amount due when the note matures is, therefore, $600 plus $120 plus $9, or $729.

At 5% simple interest $100 would amount to $160 in ten years; at annual or periodic interest it would amount to $176.20.

COMPOUND INTEREST

Compound interest is the interest that accrues by making the interest due at the close of any interval, for which the interest is made payable, a part of the interest bearing debt for the next succeeding interval. In other words the entire amount due at the end of any interval becomes the principal for the next interval. Interest may be compounded annually, semi-annually or quarterly, according to agreement. In most of the states the collection of compound interest cannot be enforced by law.

Comparison of Simple, Annual and Compound Interest

The essential difference between annual and compound interest is that in annual interest the interest for a given interval draws simple interest until the day of settlement, while in compound interest it draws compound interest until the day of settlement. The amount of $100 at 6% annual interest for ten years would be $176.20. At compound interest it would be $179.08, while at simple interest the amount would be $160. Money loaned at compound interest increases with great rapidity. A given principal will more than double itself in 12 years at 6% compound interest.

The following solution will illustrate the method of computing compound interest. Find the compound interest on $500 for 3 years, 3 months, 15 days at 4%, interest compounded annually.

SOLUTION: $500 = principal for first year
              .04
        ─────────
        $ 20 interest for first year
        $500
        ─────────
        $520 amt. for 1st yr. = principal for 2d yr.
            .04
        ─────────
        $ 20.80 interest for second year
        $520
        ─────────
        $540.80 amount for 2d yr. = principal for 3d yr.
              .04
        ─────────
        $ 21.6320 interest for third year
        $540.80
        ─────────
        $562.432 amount for 3d yr. = principal for 4th yr.
The interest on $562.432 for 3 mo. 15 da. at 4% is $6.56.
$562.43 + $6.56 = $568.99 = amt. for 3 yr. 3 mo. 15 da.
            $500
        ─────────
        $ 68.99 comp. int. for 3 yr. 3 mo. 15 da.

When compound interest is payable semi-annually or quarterly, we find the amount of the given principal for the first interval, and make it the principal for the second interval, etc. When the principal contains years, months and days, as in the problem above, we find the amount for the nearest exact interval and upon this amount compute the interest for the remaining months and days. This is then added to the last amount before subtracting the original principal.

The chief use of compound interest is among savings banks, building and loan associations, private banking houses and insurance companies. In practice a compound interest table is generally used.

A section of a compound interest table will illustrate the use. A table may be computed to any desired degree of accuracy.

The following table shows the compound amount of $1 for intervals of 1 to 10 years:

Years.	2%	2½%	3%	3½%	4%	4½%	5%
1.........	1.0200	1.0250	1.0300	1.0350	1.0400	1.0450	1.0500
2.........	1.0404	1.0506	1.0609	1.0712	1.0816	1.0920	1.1025
3.........	1.0612	1.0768	1.0927	1.1087	1.1248	1.1411	1.1576
4.........	1.0824	1.1038	1.1255	1.1475	1.1698	1.1925	1.2155
5.........	1.1040	1.1314	1.1592	1.1876	1.2166	1.2461	1.2762
6.........	1.1261	1.1596	1.1940	1.2292	1.2653	1.3022	1.3400
7.........	1.1486	1.1886	1.2298	1.2722	1.3159	1.3608	1.4007
8.........	1.1716	1.2184	1.2667	1.3168	1.3685	1.4221	1.4774
9.........	1.1950	1.2488	1.3047	1.3628	1.4233	1.4860	1.5513
10.........	1.2189	1.2800	1.3439	1.4105	1.4802	1.5529	1.6288

The amount on $1 for 4 years at 4% is $1.1698.

The amount of $300 at 4% for 4 years interest compounded annually is 1.1698 × $300; the amount of $548.60 for 2 years at 3% is 1.0609 × $548.60.

The compound amount of any given sum at a given rate

payable semi-annually, equals the compound amount of the same sum for twice the time at one-half the rate payable annually. The compound amount of any sum at a given rate payable quarterly equals the compound amount of the same sum for four times the time, at one-fourth the rate, payable annually. For example, 3 yr. 6 mo. at 4% semi-annually will yield the same compound amount as if the rate is 2% for 7 years and the interest is compounded annually. The compound amount of a given sum for 3 years at 8% compounded quarterly is the same as for 12 years at 2% compounded annually.

Applications

The subject of compound interest affords the teacher an excellent opportunity to direct attention to savings banks and the importance of these institutions. If the state laws governing savings banks are rigid enough to make them desirable institutions the teacher should direct attention to the fact that such institutions receive small as well as large deposits and pay interest upon all deposits left for a certain length of time. An appeal to economy and thrift may be made by calling attention to the rapidity with which even small deposits increase if they are regularly made and left in the bank at interest for a few years.

Ask the pupils to compute how much a boy would have in a savings bank at the end of 5 or 10 years if he deposited $1 each week and made no withdrawals, the bank paying 3% interest, compounded quarterly. Most pupils will be much surprised at the rapidity with which even small savings accumulate.

The following illustration is taken from "A Scrap Book of Elementary Mathematics," by William F. White (p. 47). It shows the enormous results obtained when compound interest is computed for long periods. The illustration will impress this fact upon the minds of pupils.

"At 3% one dollar put at interest at the beginning of the Christian era to be compounded annually would have amounted in 1906 to ($1.03)^{1906}, which, in round numbers, is $3,000,000,000,000,000,000,000,000. The amount of $1 for the same time and rate, but at simple interest, would be only $58.18."

INSURANCE

Insurance is such an important factor in modern life that the broader aspects of the subject should be taught in the schools. No attempt should be made to explain the technicalities involved in the various kinds of policies. The informational value of the subject should be emphasized rather than its mathematical content. Only the common types of property and personal insurance should be considered.

Kinds of Insurance Terms

Insurance involves a contract guaranteeing an indemnity in case of loss resulting from certain causes. When such a contract is entered into by a number of persons who agree to mutually share losses it is known as mutual insurance.

When a company is organized *as an investment* and the stockholders agree to share the profits and the losses it is known as a stock company. Sometimes the principles of the Mutual and the Stock Company are combined into a Mixed Company. In such an organization all of the earnings of the company above a specified dividend on the stock are divided. The company assuming an insurance risk is called the Insurer or the Underwriter.

The contract between the insurer and the insured is

called a policy. It states the conditions under which the guarantee against loss is made, the time the policy is to be in force, the rate, and other necessary facts.

If an insurance company guarantees against loss of property it is called a property insurance company. The principal kinds of property insurance are: fire, tornado, lightning, burglary, live stock, marine, plate glass, steam boilers, and transit. Personal insurance secures to the insured or his heirs a certain sum in case of sickness, accident or death. The principal kinds of personal insurance are life, health, and accident.

A policy may be closed or open. A closed policy is one in which the amount of the indemnity in case of loss is specified in the policy. An open policy is one in which the amount of the indemnity is to be determned after the loss.

The premium is the amount paid for the insurance. In mutual companies the cost of insurance depends largely upon the losses suffered by members of the company. In stock companies a definite premium is charged for insuring for a given time. In mutual companies the cost to the policy holder is called an assessment. The rate of insurance depends upon the nature of the risk, the face of the policy, and the period for which the policy is to be in operation. It is sometimes stated as a certain per cent of the face of the policy, but more frequently it is quoted as so much per $100.

The rates of insurance on property vary with the kinds of buildings, their location with reference to other buildings, the fire protection in the community, etc. Thus, the rate on a brick building, other things being equal, is less than on a frame structure. The rate on a livery barn is high because of the relatively great likelihood of fire in such a structure.

Fire Insurance

Fire insurance on buildings is usually written for one, three, or five years. Insurance upon the contents of a store or upon grain is frequently for a much shorter period. Fire insurance usually covers not only the loss from fire, but also from smoke and water and damage done by firemen in putting out a fire in adjoining buildings. It is customary to insure a property for about three-fourths of its actual value. Any person who has an interest in a property may insure his interest.

Some fire insurance policies contain an "average clause." When such is the case the company will in case of loss pay such a part of the loss as the policy is of the value of the property insured.

For example: when property worth $5000 is insured for one-half of its value, or $2500, the company whose policy contains an "average clause" will, in case of total loss, pay only one-half of the loss. If the "average clause" were not in the policy the entire face of the policy would be paid. Insurance companies sometimes assume a risk and then reinsure a part of the risk in other companies.

Ask the pupils to find out the different rates on various kinds of buildings in the community. Require them to account for the difference in rates. If a man constructs a building that is more nearly fire proof than other buildings of the same kind does he thereby save on his insurance premiums? Should schoolhouses and courthouses be insured? Why? Why is a property usually not insured for its full value?

Consider the details of some local fire and show what the duties of an insurance adjuster are. The teacher should secure some fire insurance policies and these should be used in the class. As far as possible the problems should be based upon local conditions.

Life Insurance

The three principal kinds of life insurance policies are the straight life, endowment and limited payment life.

The object and the principal advantages of each of these types of policies should be understood. If possible, the teacher should show the class a policy of each of these types, and the distinguishing features and relative advantages and disadvantages of each should be discussed. The advantages and disadvantages of fraternal or assessment insurance may be briefly discussed, and the chief differences between these and "old line" companies should be pointed out.

A person may insure his own life or that of any person in whom he has a pecuniary interest, or upon whom he depends for support. The necessity for a physical examination of the applicant for insurance and for the numerous facts of his family history should be made evident.

Pupils will be much interested in comparing the rate of increase in cost of insurance as a man becomes older. Pupils are especially interested in a table which shows the expectancy of life.

CARLISLE TABLE OF EXPECTANCY OF LIFE

Age.	Expectancy in years.	Age.	Expectancy in years.	Age.	Expectancy in years.	Age.	Expectancy in years.
0	38.72	26	37.14	52	19.68	78	6.12
1	44.68	27	36.41	53	18.97	79	5.80
2	47.55	28	35.69	54	18.28	80	5.51
3	49.82	29	35.00	55	17.58	81	5.21
4	50.76	30	34.34	56	16.89	82	4.93
5	51.25	31	33.68	57	16.21	83	4.65
6	51.17	32	33.03	58	15.55	84	4.39
7	50.80	33	32.36	59	14.92	85	4.12

CARLISLE TABLE OF EXPECTANCY OF LIFE—Continued

Age.	Expectancy in years.	Age.	Expectancy in years.	Age.	Expectancy in years.	Age.	Expectancy in years.
8	50.24	34	31.68	60	14.34	86	3.90
9	49.57	35	31.00	61	13.82	87	3.71
10	48.82	36	30.32	62	13.31	88	3.59
11	48.04	37	29.64	63	12.81	89	3.47
12	47.27	38	28.96	64	12.30	90	3.28
13	46.51	39	28.28	65	11.79	91	3.26
14	45.75	40	27.61	66	11.27	92	3.37
15	45.00	41	26.97	67	10.75	93	3.48
16	44.27	42	26.34	68	10.23	94	3.53
17	43.57	43	25.71	69	9.70	95	3.53
18	42.87	44	25.09	70	9.18	96	3.46
19	42.17	45	24.46	71	8.65	97	3.28
20	41.46	46	23.82	72	8.16	98	3.07
21	40.75	47	23.17	73	7.72	99	2.77
22	40.04	48	22.50	74	7.33	100	2.28
23	39.31	49	21.81	75	7.01	101	1.79
24	38.59	50	21.11	76	6.69	102	1.30
25	37.86	51	20.39	77	6.40	103	0.83

The expectancy of life of a man who is 22 years of age is 40.04 years; that of a man who is 46 years of age is 23.82 years.

The premium charged by a given life insurance company for a given policy is determined by three considerations: (1) The age of the insured. (2) The expenses of managing the company. (3) The rate of interest that the company can earn upon the premium received.

State laws differ greatly in the extent to which they protect policy holders, and this fact should be mentioned in discussing the subject of insurance.

Illustrations

Suppose that Mr. A, who is 25 years old, takes out an ordinary life policy for $1000, for which he pays an annual premium of $21.35. After paying four annual premiums, or $85.40, A dies and the company pays his beneficiary $1000.

Pupils may be puzzled to know how the company can afford to do this. The expectation of life for a man 25 years of age is 37.86 years, and many who take insurance when they are 25 years of age live more than 37.86 years. There are many who take out a policy and after carrying it for a few years drop it and so receive nothing. If A had lived 50 years after taking out his policy and had paid his annual premiums he would have paid in $1067.50. The insurance company would have put each sum on interest when it was paid and the entire sum at compound interest would probably have amounted to more than $2500. A might have invested his money at a higher rate than the insurance company did, but he was willing to pay $21.35 annually to an insurance company for assuming the risk on his life.

Similar considerations may be discussed for the endowment and the limited payment life policies. The reason why the limited payment life costs more than the straight life and why the endowment costs more than the other two should be clear to the pupil.

The text-book problems on insurance involve no mathematical difficulties, and if the pupil understands the general features of the subject he should have no difficulty in solving the problems.

TAXES AND REVENUE

Unlike some of the applications of percentage, the subject of taxes is one in which most of the homes represented by the pupils are directly interested. Once or twice each year the payment of taxes becomes a very real matter and the entire subject can be made interesting to the pupils if most of the problems are based on local conditions. A detailed discussion of the subject is a part of civics rather than of arithmetic.

Why Taxes Are Necessary

The first essential is that the pupils should understand why the levying and collecting of taxes is necessary. It should be explained that a tax is any money levied by a government,—national, state, county, city,—to defray all or a part of its expenses. Pupils should be asked to name several items of expense of the nation, state, county, township, and city; such as the salaries of all state officials, erection and maintenance of public hospitals, asylums and prisons, state schools and public buildings. The county must raise money to defray the expense incurred in paying the salaries of county officials, the erection and maintenance of the court house and other county buildings, the building and repairing of roads, etc. The city must levy a tax to maintain the public schools, to provide proper sewerage and other sanitary conditions, to provide police and fire protection, to pay the salaries of city officials, etc. The smaller units of government must levy tax to build and repair roads and bridges and to defray other items of public expense. Ask the pupils to compare the taxes which a man pays each year with what he would have to pay to secure equivalent service and protection if no taxes were levied. If taxes were not levied he would have to pay for the education of his own children, and this item alone would, in most cases, be larger than his annual taxes. He would also need to provide for protection from fire and theft. The pupils should see that a just levy of taxes is not only wise but economical for the individual citizen.

Kinds of Tax

For purposes of taxation there are two kinds of property tax,—real estate and personal. Real estate comprises property that is not easily moved, such as land, mines, quarries, buildings, railroads, and forests. Personal property is easily moved, such as money, stocks, bonds, live stock, and household goods. Separate assessments are made for each kind of property, and in some localities the personal and real taxes are due on different dates. Pupils should be asked to explain why the rates of city and county taxes in a given locality differ. What protection or convenience does the man who resides in the city have that his friend in the country does not have? Is it reasonable that the resident of the city should be assessed for this protection and convenience?

Besides the property tax just mentioned there are several other kinds of tax. In many states a poll tax of a specified sum is levied on all male citizens over twenty-one years of age. Franchises which are granted by municipalities or counties are often subject to a special tax. In many localities a special dog tax is levied. A special tax is also levied in some states on all inheritances over a certain prescribed minimum.

Method of Levying and Collecting Taxes

The method of levying and collecting taxes varies in the different states and a detailed consideration of the various methods would be too long for this book, but the teacher should not fail to ascertain the method used in the state and in the county in which he is teaching and to familiarize the pupil with it. Such a study should acquaint the pupil with the meaning and use of the following terms: valuation, assessment, assessor, board of review, board of equalization, tax collector, and delinquent taxes.

How Tax Rates Are Determined

The method by which the rate for a given kind of tax is determined should be clearly understood. The state legislature determines the amount to be spent by the state. The assessed value of the property of the state is ascertained and from these two items the rate of the state tax is determined. For illustration: if the state legislature authorizes the expenditure of $18,000,000 and the property valuation in the state is $3,000,000,000, the rate of the state tax will be .006. Tax rates are usually expressed as so many mills on each dollar of valuation. In the illustration just given the state tax would be 6 mills on each dollar. The rate for school, road and bridge and other taxes in a community is determined in the same general manner. The pupil should be told what body is authorized to determine the amount of each of the above taxes.

In many localities it is the custom to assess property at $\frac{1}{2}$ or $\frac{2}{3}$ of its real value.

The time and labor of computing taxes are reduced by the use of tables from which the tax on various sums at specified rates may be easily found.

Base Problems on Local Conditions

Most teachers can obtain an assessment blank and several tax receipts for class use. Problems based on local levies and assessments are of much more interest to the pupil than most of the problems of the text-book. Details in regard to state revenue laws may be secured by addressing the secretary of state.

Duties and Revenues

The subject of national duties and revenues is closely related to that of taxes and may be presented in a similar manner. Citizens are not taxed directly for the support of the national government. Among the chief items of national expense may be mentioned the maintenance of the Army and the Navy, Diplomatic and Consular Service, Indians, Internal Improvements, Interest on the Public Debt, Erection and Maintenance of Public Buildings, Salaries of government officials, Pensions, etc. The expenses of the National Government are defrayed by the receipts from duties and customs, internal revenues (liquors, tobacco, etc.), sale of public lands, income and corporation taxes, etc.

The study of revenues introduces the question of specific and ad valorem duties, the tariff and kindred subjects. The teacher must use his discretion as to how fully these subjects should be considered in a class. The mathematics involved in the problems of taxes and national revenues is so simple that illustrative problems are not necessary. The entire subject should acquaint the pupil with local and national methods of defraying necessary expenses. This involves a study of the elements of civics and of economics rather than of arithmetic. The mathematics involved should be considered as subordinate to the other phases of the subject.

CHAPTER XVII

BANKING; CORPORATIONS; BUSINESS PRACTICE

The modern bank is a complex institution; it is not the duty of the school to acquaint pupils with all of the technicalities involved in banking. Pupils should know the function of a bank in modern commercial life; the principal duties of the book-keepers and the cashiers; how to deposit money and to make out and endorse checks.

The Chief Functions of a Bank

1. To receive money on deposit and pay checks drawn on deposits.
2. To lend money on promissory notes.
3. To sell drafts on other banks and collect drafts on persons.
4. To issue its own promissory notes, which serve as currency (only National Banks can do this in the United States.)

The School Bank

Problems involving banking should be made realistic to the pupils. The nearer the approach to actual business the greater will be the interest of many of the pupils. The school may have a bank in which certain pupils act as cashier, paying teller, receiving teller, bookkeeper, etc. Inexpensive paper money, deposit slips, and checks may be secured. More than ninety per cent of the business of the country is carried on by checks and drafts, but few pupils have seen either. In some localities it will be possible to induce a local banker to talk to the pupils on such subjects as the clearing house.

Negotiable Paper

Pupils should become familiar with the principal forms of promissory notes, bills of exchange, and checks. These instruments are an important factor in business transactions, passing from hand to hand as a substitute for money.

No exact form need be followed in order to make a contract a negotiable instrument, but custom has prescribed forms that are very generally used. A negotiable instrument, whether it is a promissory note, bill of exchange or check, must be:

(a) In writing. (No oral contract is negotiable).

(b) Properly signed. (The name is not necessary; any mark intended to serve as a signature will suffice).

(c) Negotiable in form. (If made payable to a particular person or persons it is not negotiable).

(d) Payable in money only. (The value is then definite and certain).

(e) The amount must be definitely stated.

(f) It must be payable absolutely. (Not upon certain conditions).

(g) To the order of a designated payee or bearer. (To a person or persons who can be identified when the note matures).

(h) At a certain time. (Not at a time contingent upon some other event, but at a time sure to arise).

Promissory Notes

The party who makes the note and whose promise is stated is called the *maker*. The party to whom the promise is made is called the *payee*. If the note is without interest this fact may be stated in the note or the words "with interest" may be omitted.

The *payee* of a negotiable note may transfer his rights to another by writing his name upon the back of the note. This is called an *indorsement*, and the person to whom the note is transferred is called the *indorsee*.

$1800.   Urbana, Illinois,
                                      September 28, 1913.

Sixty days after date I promise to pay to the order of James Riley, Eighteen Hundred Dollars, without interest. Value received.

                                            JOHN BAKER.

Ask the pupil who is the maker of the above note? Who is the payee? When is the note due? Who pays the note when it matures? To whom? How much must be paid when the note matures? Who gets the check after it is paid? Why are the words "value received" put in a note? Is the above note negotiable? Suppose the payee wishes to sell the note before it matures; how should he indorse it?

The endorsement in the illustration is called an endorsement in full. If the note is endorsed by the words "pay to William Jones only" it cannot be transferred again. This is called a "restricted endorsement." The holder of the note may simply write his name across the back of it. This is called an endorsement in blank, and makes the note payable to bearer. The question of liability incurred by the indorser of a note should be discussed. The extent to which the indorser of a note limits his liability by using the words "without recourse" should be considered.

There are numerous kinds of notes and drafts involving more or less technicalities. The teacher should acquaint the pupils with only the simpler forms in common use. In treating the subject of negotiable paper a good commercial arithmetic will be of value to the teacher.

Pay to the order of
WILLIAM JONES.
JAMES RILEY.

When money is borrowed from a bank the security given may be (a) real estate, (b) collateral, (c) personal. These should be discussed in the class and given practical setting by use in the *school bank*.

Discounting Notes

Suppose that James Riley, the payee of the note on page 273, sells the note to a bank on October 23. It is evident that the bank could not afford to pay $1800 for it because that amount will not be paid to the holder of the note until it is due,—November 27; the bank would lose the interest on $1800 from October 23 to November 27.

The interval between the date a note is discounted and the date of maturity is called ''the term of discount.'' The difference between the value of a note at maturity and the amount paid for the note when it is sold is called discount.

A note that does not bear interest matures at its face value. If it is interest-bearing from date, the interest which will have been earned at the date of maturity must be added to the face of the note to determine its maturity value.

The first step in calculating discount on a note is to find the maturity value since the purchaser must know this before he can determine how much he can afford to pay for it.

In computing the term of discount banks usually count the exact number of days. In some states days of grace are still permitted. If a note is due a certain number of months after date it will mature on the same day of the month as the date of the note.

For convenience the following table is often used by bankers to find the days intervening between dates.

DISCOUNT TABLE

From any day	January	February	March	April	May	June	July	August	September	October	November	December
To the same day of the next												
January	365	31	59	90	120	151	181	212	243	273	304	334
February	33	365	28	59	89	120	150	181	212	242	273	303
March	306	337	365	31	61	92	122	153	184	214	245	275
April	275	306	334	365	30	61	91	122	153	183	214	244
May	245	276	304	335	365	31	61	92	122	153	184	214
June	214	245	273	304	334	365	30	61	92	122	153	183
July	184	215	243	274	304	335	365	31	62	92	123	153
August	153	184	212	243	273	304	334	365	31	61	92	122
September	122	153	181	212	242	273	303	334	365	30	61	91
October	92	123	151	182	212	243	273	304	335	365	31	61
November	61	92	120	151	181	212	242	273	304	334	365	30
December	31	62	90	121	151	182	212	243	274	304	335	365

The exact number of days from any day of one month to the same day of another month, within a year is found by starting at the name of the first month in the left hand column and following across to the column of the last named month. Thus to find the number of days from

April 18 to August 24, we note that it is 122 days from
April 18 to August 18, and then we add 6 days (August
18 to August 24) which gives 128 days as the exact time.

Suppose we wish to find the bank discount on the note
given on page 273, if the note is discounted on October 23
at 6%.

$1800 = maturity value of the note.   (Future worth).

Time from October 22 to November 27 is 36 days, this
is the term of discount.

Interest on $1800 for 36 days at 6% is $10.80, this is
the bank discount.

The seller of the note would receive $1800 – $10.80, which
is $1789.20.

If the preceding note had been drawn with interest at
6% the maturity value of the note would have been $1800
plus the interest on $1800 at 6% for the time of the note
(60 days.)   This would have been the basis for computing
the bank discount.

Partial Payments

The subject of partial payments is no longer of great
value because of changes that have been made in the form
of notes.   The modern bank is rendering the subject obsolete.   The numerous state rules should not be taught.

Methods of Transmitting Money

Some attention should be devoted to the various ways of
sending money from one locality to another.   The advantages and disadvantages of each should be briefly discussed.

Debts between men in the same community are usually
paid by means of checks or of actual currency.   Debts
between persons in different communities may be dis-
charged by any of the means enumerated below:

1. Sending actual currency or stamps.
2. Check.
3. Bank draft.
4. Registered letter.
5. Postal money order.
6. Express money order.
7. Telegraph money orders.

Very little attention should be given to the technicalities involved in the subject of exchange.

Bookkeeping

Pupils should be taught how to keep simple accounts. The average man or woman finds but little occasion for any elaborate form. Most business men prefer to teach their own method to one who will have charge of their books.

Pupils should, however, be taught how to keep a neat and accurate record of the expenses of the home, the farm, the store, and the like.

Stocks and Bonds

The subject of stocks and bonds is generally considered to be the most difficult application of percentage. The difficulty is due, however, not to the mathematics involved, but to the fact that the language of the subject is frequently unfamiliar to the teacher and is usually unfamiliar to the pupil.

Organization of a Corporation

The best way to approach the subject in order to make it real to the pupils is to let the pupils organize a corporation. When such an organization is completed the pupil will have a better knowledge of the meaning of the technical terms of stocks and bonds than he would have had from a study of the terms in a more formal manner. Corporations are assuming control of many business enterprises and are replacing the individual in larger business ventures.

Before a corporation can be organized it must secure a commission from the secretary of state. The persons commissioned by the secretary of state are permitted to take subscriptions for the stock. After the stock is subscribed a meeting is held for the purpose of effecting a temporary organization. A constitution is adopted and the officers provided for in the constitution are elected. The name and purpose of the organization, its location, the number of shares and par value of each share, the officers and the time for holding regular meetings of the stockholders are determined upon and a statement of these facts is filed with the secretary of state. A charter is then issued and the organization is completed. The stock previously subscribed for is sold. The directors are empowered to buy, sell, contract debts, etc.

Suppose that the pupils organize a company with a capital stock of $100,000, the par value of each share being $100. (The par value may be $500 or $50, or $10, or any other convenient amount.) The pupil who buys 20 shares of the stock will pay $2000 for it, if the stock sells at par value. For each share that one owns he is entitled to one vote at all meetings of the stockholders. When one pays for his stock he receives a "certificate of stock," stating the number of shares bought, the par value of each share, etc. The owner of stock is entitled to participate in the profits in proportion to the number of shares owned.

Corporations often issue both preferred and common stock. A preferred stock certificate states that dividends up to a certain limit, usually 4% to 7%, are to be paid before any dividends are paid on common stock. Holders of preferred stock have first chance at dividends, but their dividends are limited.

Suppose that at the end of the first year the corporation organized by the pupils has cleared $12,000 above all expenses. The directors may decide to declare this as a dividend, or all or a part of it may be put into a "reserve fund." If it is declared as a dividend and if the holders of preferred stock are entitled to 6%, this must be paid before any dividend is computed for the holders of common stock. Let us suppose that $30,000 of the entire stock ($100,000) is preferred stock. The holders of this stock are entitled to 6% of $30,000, or $1800. A dividend of $6 is paid on each share. The holders of the common stock are entitled to the balance of earnings, or $10,200. This dividend will be divided among the holders of $70,000 worth of stock. Each holder of common stock will, therefore receive 14$\frac{4}{7}$% dividend on each share, since this is higher than the current rate of interest, the common stock will sell above par, that is, at a premium.

Suppose that at the end of the second year the corporation has cleared but $1800. If dividends are declared the holders of preferred stock are entitled to this entire amount, since it is just 6% of the face value of the preferred stock. The holders of common stock would receive no dividends and their stock would likely sell below par. In case the net earnings of the corporation are not sufficient to pay 6% dividends on the preferred stock, the earnings of the corporation, if declared as dividends, will be divided pro rata among the holders of preferred stock.

Bonds

Corporations may borrow money, pledging their property as security in the same way as individuals. When a large amount of money is to be raised it is usually done by issuing *bonds*. A bond is a mortgage note, upon which the corporation agrees to pay interest, and to the payment of which the entire property and business of the corporation is pledged.

Bonds may be issued by national and state governments, counties, townships, cities, school districts, etc. They are declared by law to be "bodies corporate."

Bonds made payable to the owner or his assignee are called registered bonds; bonds made payable to bearer are termed coupon bonds. In such bonds the interest is provided for in attached notes or "coupons." In the case of registered bonds the interest is sent directly to the owner. The holder of a coupon bond must surrender a coupon when each interest payment is made.

If a bond pays more than the current rate of interest and is safe, it usually sells at a premium.

The price of many stocks and bonds is quoted in the daily papers. Market values often fluctuate widely and an intimate knowledge of corporations and market conditions is necessary if one expects to invest wisely. A broker is assumed to know the stocks and bond market and to be able to advise his clients when to buy or to sell stocks and bonds. The fee paid him for his services is called commission or brokerage, it is usually about $\frac{1}{8}$%. It should be remembered that brokerage is always a per cent of the par value whether stock and bonds are bought or sold. The technicalities of the broker's office are not of value to most citizens and need not be taught in the schools.

Illustrative Problems

Many of the problems in stocks and bonds should be based upon the market quotations given in the daily papers.

Illustrative problems :.

A broker sells 160 shares of stock at 90; brokerage $\frac{1}{8}\%$. What should his principal receive?

1. $90 - $\frac{1}{8}$ = proceeds from each share.

   $160 \times \$89.875 = \$14,380 =$ proceeds from 160 shares.

2. What sum must be invested in 5% bonds at 110, without brokerage, to yield an annual income of $450?

Each bond yields $5 dividend annually (5% of par value).   •

To yield $450 the number of bonds must be $450 ÷ $5, which is 90.

Since each bond cost $110, the entire cost was $90 \times \$110 = \$9900$.

3. What per cent does an investor make on stock that pays a dividend of 5% if he buys it at 80, without brokerage?

$5 = income on one share.

$80 = cost of one share.

The return on the investment is whatever per cent $5 (the dividend) is of $80 (the amount invested.) $5 is $6\frac{1}{4}\%$ of $80.

4. What is the quotation price of 7% stock that yields an income of 10%, no brokerage?

$7 = income on each share.

$7 = 10% of the amount invested in one share, therefore, the amount invested is $70.

CHAPTER XVIII

THE METRIC SYSTEM

Essentials of a Perfect System

Any perfect system of weights and measures must have two characteristics. The units must be accurately defined and invariable and there must be a uniform ratio between consecutive units in any given table. The early systems of weights and measures had neither of these essentials. The units were not accurately defined and the scale of relation was variable. The necessity for accurately defined units became more imperative as commerce and industry developed, and eventually the units were defined with scientific accuracy.

Science improved the old system of weights and measures by imparting to its units the exactness and uniformity which is the first essential, but it did not establish the simple scale of relationship between the units, which is the second essential.

The lack of uniformity in the ratio between consecutive units of the English system may be illustrated by a consideration of the following tables:

$$12 \text{ inches} = 1 \text{ foot}$$
$$3 \text{ feet} = 1 \text{ yard}$$
$$5\tfrac{1}{2} \text{ yards (or } 16\tfrac{1}{2} \text{ feet)} = 1 \text{ rod}$$
$$320 \text{ rods (or } 5280 \text{ feet)} = 1 \text{ mile}$$
$$16 \text{ ounces} = 1 \text{ pound}$$
$$100 \text{ pounds} = 1 \text{ hundred weight}$$
$$20 \text{ hundred weight} = 1 \text{ ton}$$

Not only is there a lack of uniformity of ratio between the consecutive units throughout a given table, but the same ratio is rarely maintained for three consecutive units. In linear measure the numbers expressing the relationship between consecutive units are 12, 3, 5½, and 320. In surface measure the ratios are 144, 9, 30¼, 160, and 640.

In avoirdupois weight the ratios are 16, 100, and 20.

It is evident that the English system of weights and measures is lacking in the second essential of a perfect system,—a uniform scale of relationship between the consecutive units of a given table.

Origin of Metric System

The Metric system is the result of an attempt by the people of France to devise a perfect system of weights and measures,—a system that would have both of the essentials that have been mentioned. It was first suggested in 1528, but it was not worked out until 1790. The Metric system was the product of the minds of five of the most eminent mathematicians of France. Its history is very interesting and will repay study. The name of the system is derived from the French word *metre,* meaning "to measure." The meter was defined to be one ten-millionth of the distance from the equator to the pole, and this was taken as the standard of linear measure. All other units of the system were correlated directly with the meter. It was later discovered that a slight mistake was made in computing the length of the meter, but this error does not render the system less valuable.

Present Status of Metric System

The enlightened judgment of educated people pronounces the system to be the best the world has yet seen. It is endorsed with enthusiasm by those who have studied it, and many predict that it will ultimately triumph over the English system and will become universal among civilized nations. The persistence of custom will delay its ultimate triumph for many years, but the best must eventually triumph. Since its adoption in France in 1840 it has spread with great rapidity. It is to-day either obligatory or permissive in every civilized country of the world. It is in general use among all civilized nations except England and the United States. Instruction in it is obligatory in all English schools. It has been the legal system in the United States since 1866; its use is required in some of the departments of our government, and is authorized in others.

The Future of the Metric System

We live in an age of international activities. The tendencies of the day are towards uniformity of language and customs. Commerce demands that articles for export and for import shall be measured by units that are universally understood among civilized nations. The complex business activities of the day make the use of a simple system imperative. One advocate of the metric system calls attention to the fact that "there is no nation to-day having a decimal system of weights and measures or currency where there is a suggestion of a change, and no nation not having a decimal system where there is not constant agitation in favor of a decimal system. Whenever a civilized nation has changed its units the change has always been decimal."[1] Another ardent advocate of the metric system states the following: "Considered merely as a labor-saving machine

[1] Hon. James E. Southard, "School, Science and Mathematics," 1905, pp. 653-657.

it is a new power offered to man incomparably greater than that which he has acquired by the agency which he has given to steam. It is in design the greatest invention of human ingenuity since that of printing.'' [2]

There are two objections advanced to the general adoption of the metric system in the United States. The first is a real objection, but it is not so formidable as is supposed; i. e., the cost of changing tools to correspond to the metric units and the training of the laborer to think and work in terms of these units. A sudden change to the metric system would be attended with confusion, but no such change will take place, because the adoption of the new units will be made gradually. Our money system is now on a decimal basis. In recent years several of the large manufacturing concerns in the United States have adopted the metric units for exclusive use in shops, and the management report little or no confusion as a result of the change. The second great objection urged against the adoption of the metric system is the fact that in the matter of weights and measures we are strongly bound by the chains of tradition. The chain of tradition is weakening, however, and the metric system is gaining ardent advocates at the expense of the English system.

It is desirable that teachers should appreciate the merits of the metric system. No teacher can afford to be ignorant of a system that has been so generally adopted by civilized nations and is the basis for all scientific work. A well-educated person to-day should know the metric units in common use. The English system should receive much greater emphasis than the metric system, for it is the system that most pupils will use in their daily activities.

[2] Hallock and Wade, ''Evolution of Weights and Measures and the Metric System,'' pp. 116-118.

Where the Metric System Should Be Taught

The metric system should usually be taught in the seventh or eighth grades, and its superiority should be emphasized. It is better to teach the system in the seventh or eighth grade than earlier, because it can be closely correlated with the work in science and with the industrial and commercial activities which are usually emphasized in these grades.

Suggestions on the Teaching of the Metric System

The metric units in common use should be actually in hand. The best results cannot be secured unless the pupils see and handle the measures and they should be studied as metric units, without regard to their nearest English equivalents. Pupils should be trained to think in terms of metric units. The power to estimate accurately in terms of these units may be developed through drill. Ask the pupils to hold two fingers one, two, or three centimeters apart. Estimate the linear and surface measure of numerous common objects. Most of the work in the metric system should be oral. After the units are well understood and some facility has been acquired in their use, a brief study of the equivalents of the English system may be made. (A table of equivalents is given at the close of this chapter.)

The abbreviations for the metric units are not uniform even in France. Square meter is written as qm, or m², or □M. Each compound name in the metric system is accented on the first syllable, thus: kil′ometer; mil′limeter.

The prefixes must be thoroughly learned if one is to master the metric system. The pupil must know that "kilo" always means 1,000, whether it is prefixed to the unit, meter, liter, or gram. A kil'ogram is 1,000 grams; a kil'ometer is 1,000 meters; a kil'oliter is 1,000 liters. Similarly, "milli" when prefixed to any units means .001 of that unit. Thus, millimeter means .001 of a meter. The following mnemonics may help the beginner to associate the new terms with names which are familiar. The word "meter" is familiar to the pupil in such terms as "water meter," "gas meter," and "cyclometer." The term always means "a measure." Deci, meaning .1, may be associated with a dime, 0.1 of a dollar. Centi, meaning .01, may be associated with cent, .01 of a dollar. Milli, meaning .001, may be associated with mill, .001 of a dollar. If the term dekagram, a ten-sided figure, is familiar to the pupil, he will more easily remember that deka means ten. Myria is easily remembered as a large multiplier by associating it with the word myriad.

The prefixes should be drilled upon until they are mastered and the pupil is able to translate back and forth in the table. The prefixes which indicate multiples of the unit are derived from Greek, and those which indicate the decimal divisions of the unit are from Latin.

The remarkable simplicity of the metric system and the ease with which reductions may be made should be emphasized. For example, compare the ease with which 487.26 kilometers may be reduced to meters with a similar reduction in the English system. Reductions in the metric system involve only the moving of the decimal point; thus the subject gives an opportunity to review decimals. Since 1,000 liters equal 1 cubic meter, the capacity of bins and tanks may be computed with much greater ease than by use of the English units.

The five-cent piece may be used to illustrate some of the metric units. Its diameter is 2 centimeters; its thickness is 2 millimeters, and its weight is 5 grams.

The definitions of meter, liter, and gram should be learned.

A meter is .0000001 of the distance from the equator to the pole. It is the unit of length and is the base of the entire metric system.

A liter (pronounced "lēter") is the capacity of a cube .1 meter on an edge. It is the unit of capacity.

A gram is the weight of a cube of distilled water .01 of a meter on an edge. It is the unit of weight.

Prefix	Meaning	Illustration	Meaning
myria	10,000	myriameter	10,000 meters
kilo	1,000	kilogram	1,000 grams
hekto	100	hektogram	100 grams
deka	10	dekameter	10 meters
deci	0.1	deciliter	0.1 liter
centi	0.01	centimeter	0.01 meter
milli	0.001	millimeter	0.001 meter
mikro	0.000001	mikrometer	0.000001 meter

Table of Equivalents

A meter	= 39.37 inches = 3¼ ft. nearly = 3 ft. 3 inches, ⅜ inch, nearly.
A Liter	= 1 quart nearly
A kilogram	= 2.2 pounds nearly
A gram	= 15.43 grains
A hektare	= 2.47 acres

The meter is used for dry goods, merchandise, engineering, construction, and other purposes where the yard and foot are used in the English system.

The centimeter and millimeter are used instead of the inch and its fractions in machine construction and similar work.

The centimeter is used in expressing sizes of paper and books.

Any quantity consisting of several denominations may be treated as an integer and a decimal, the decimal point separating the unit and its divisions. For example, 1420.25 meters is not read 1 kilometer, 4 hektometers, 2 dekameters, 2 decimeters, and 5 centimeters, but is read 1,420 meters and 25 centimeters, just as $1420.25 is read $1420 and 25 cents.

Unit	Where Employed
Megameter	Astronomy.
Myriameter	Geography
Kilometer	Distances in general
Hektometer	Artillery
Dekameter	Surveying
Decimeter	⎰ Commerce
Centimeter	⎨ Industry and
Millimeter	⎱ Science
Mikron	⎰ Metrology
Millimikron	⎨ Spectroscopy
	⎱ Microscopy

CHAPTER XIX

INVOLUTION AND EVOLUTION

The subjects of Involution and Evolution were treated in some of the early texts, but most of the medieval arithmetics did not discuss the subject of involution extensively. The modern treatment of this subject was introduced from algebra. Evolution, as studied in the grades, presupposes but little knowledge of involution, and the movement towards the elimination of topics will probably reduce the treatment of involution.

Neither involution nor evolution is frequently used in business. A knowledge of squares and cubes, of square roots and cube roots, is necessary in certain scientific work, but the scientist and the engineer use tables or the slide rule to secure the results desired in the computation.

Cube root has been eliminated from most courses in arithmetic. A recent report indicates that the subject is still taught in about 28% of the elementary schools, but it will no doubt be eliminated from many of these schools within the next few years. The study of cube root is being postponed until the pupil encounters the subject in algebra, or is omitted entirely. Square root is taught in most schools in the eighth grade. The chief reason for its retention in the course is the fact that it is needed in the mensuration of many geometrical forms. For example, a knowledge of square root is necessary in finding the diagonal of of a rectangle or of a cube, the altitude and area of an isosceles triangle, and the slant height of a pyramid or cone. If square root is applied to the solution of some real problems, pupils will be interested in it.

The Terms Used

The technical terms of involution and of evolution should be thoroughly understood. The meaning of power, exponent, square and square root, radical sign and digit should be familiar to the pupil. Much time in computation will be saved if the pupil is required to learn the squares of all integers from 1 to 20 or 25, and the cubes from 1 to 10. Pupils should be able to state at sight the square roots of such numbers as the following:

64   400   900   81   36   3600

Square Root by Factoring

The first examples worked in square root should be by factoring. The fact should be emphasized that we seek to find two equal numbers whose product is the given number. $2025 = 5 \times 5 \times 9 \times 9$, or $5^2 \times 9^2$; therefore the square root of $2025 = 5 \times 9 = 45$. Similarly, $324 = 2 \times 2 \times 9 \times 9$, or $2^2 \times 9^2$; therefore the square root of $324 = 2 \times 9 = 18$. Similarly, $53361 = 3 \times 3 \times 7 \times 7 \times 11 \times 11$, or $3^2 \times 7^2 \times 11^2$; therefore the square root of $53361 = 3 \times 7 \times 11 = 231$.

Sometimes we wish to obtain the square root of a number not readily factored, or the approximate square root of a number that is not a perfect square. If we wished to find the square root of 5329, or the approximate square root of 3, the desired result could not be conveniently found by factoring.

Other Methods for Extracting Square Root

There are two general methods for explaining the extraction of the square root of such numbers. One of these is known as the *algebraic* and the other as the *geometric* method. One of the methods is analytic and the other is synthetic. Text-book writers are not agreed as to which method should be used in explaining the process of evolution, and several of the best books give both methods. Occasionally a teacher advocates the teaching of square root as a purely mechanical process, with no attempt to justify the procedure to the mind of the pupil. Such a point of view is contrary to the best educational theories of the day. It is not assumed that the explanation of square root by either the analytic or the synthetic method will so impress itself upon the mind of every pupil that each one will be able to explain the theory involved in a given problem. Pupils should understand each step of the process, but the object is not to enable the pupil to repeat the explanation in a more or less mechanical way; it is to justify the various steps to his mind. After this has been done the pupil should be required to work numerous examples until he has acquired facility in the process. Pupils should be required to formulate a rule for square root and to memorize it.

Algebraic Method. This method is applicable to the extraction of any desired root, but no root higher than the third is necessary in arithmetic.

The square of 45 may be found as follows:

$$
\begin{array}{ll}
(45)^2 = (40+5)^2 & \text{Similarly, } (97)^2 = (90+7)^2 \\
\quad 40 + 5 & \quad 90 + 7 \\
\quad 40 + 5 & \quad 90 + 7 \\
\hline
40^2 + \ (40 \times 5) & 90^2 + \ (90 \times 7) \\
\quad + \ (40 \times 5) + 5^2 & \quad + \ (90 \times 7) + 7^2 \\
\hline
40^2 + 2\,(40 \times 5) + 5^2 & 90^2 + 2\,(90 \times 7) + 7^2
\end{array}
$$

Every number composed of two or more digits may be regarded as composed of tens and units. Thus, 45 equals 4 tens plus 5 units; 97 equals 9 tens plus 7 units. 434 equals 43 tens plus 4 units. By working several examples similar to the above, we find that the square of any number will equal the square of the tens, plus twice the tens by the units, plus the square of the units. If t represents the number of tens in any given number and u represents the number of units, we may say that the square of the number equals $t^2 + 2tu + u^2$.

Since the square of the smallest number of one digit contains one digit, and the square of the largest number of. one digit contains two digits, the square of any number of one digit must contain either one or two digits. Similarly, since the square of 10, the smallest number of two digits, is 100, and the square of 99, the largest number of two digits, is 9801, we conclude that the square of a number of two digits must contain three or four digits. This may be extended to larger numbers. If any integral number is divided into groups of two digits each, from the right to the left, the number of digits in the root will be the same as the number of groups of digits. The last group to the left may contain one or two digits. Thus the square root of $\overline{18}\ \overline{14}\ \overline{76}$ contains three digits. The square root of $\overline{169}$ contains two digits.

Required to find the square root of 2025.

If 2025 is a perfect square, its square root must contain two digits, since the number itself contains four digits. The largest square in 20 is 16, and the square root of 16 is 4. The ten's digit of the root is therefore 4. The square of 4 tens, or 40, is 1600. 425 is composed of two times the tens, times the units, plus the square

$$
\begin{array}{r}
4\ \ 5 \\
\hline
2025 \\
1600 \\
\hline
80\,|\,425 \\
5\,|\,425 \\
\hline
85
\end{array}
$$

of the units. We do not know the square of the units, but we know that it is small compared with two times the tens times the units. We may therefore think of 425 as composed approximately of two factors; one of these factors (two times tens) is known, and the other is not known. We therefore divide 425 by two times tens in order to find the units, which is the other factor. In this division we remember that 425 is somewhat larger than the product of these factors, hence we make a slight allowance for this excess. In the problem above, two times the tens equals 2×40, or 80. 80 is contained in 425 five times. The units digit of the root is 5. The 5 is added to the 80 and this result is then multiplied by 5, because we may regard two times the tens times the units plus the square of the units as (two times the ten plus the units) times units.

In practice we may omit the zeroes in the square of 40, also the zero of 80. We may then annex the 5 to the 8, and the work appears as follows:

$$
\begin{array}{r}
4\ 5 \\
\hline
20\ \overline{25} \\
16 \\
85\ \overline{)\ 4\ 25} \\
4\ 25
\end{array}
$$

The method just explained may be used in finding the square root of any perfect square or the approximate square root of a number whose exact square root cannot be found. If the given number contains more than two groups of two digits each, we may think of the part of the root found as so many tens with reference to the next digit to be found, and the process is the same as before. This should be clear from the illustration that follows:

Required to find the square root of 107584.

$$
\begin{array}{r}
3\ 2\ 8 \\
\hline
10\ \overline{75}\ \overline{84} \\
9 \\
\hline
62\ |\ 1\ 75 \\
|\ 1\ 24 \\
\hline
648\ \ \ \ |51\ 84 \\
|5184
\end{array}
$$

The square root of 107584 is therefore 328.

The rule for grouping the digits of a decimal whose square root is required may be easily developed by a method similar to the one followed in determining how to group the digits of an integral number. Since the square of tenths is hundredths, and the square of hundredths is ten thousandths, we conclude that in extracting the root of a number containing a decimal fraction we must begin at the decimal point and group the digits in twos from the left to the right. Thus, $.\overline{57}\ \overline{46}\ \overline{28}$; a zero may be annexed to complete the number of digits in the last group. Thus, $.\overline{38}\ \overline{75}\ \overline{2} = .\overline{38}\ \overline{75}\ \overline{20}$. With the exception of this difference in grouping the digits, the extraction of square root is practically the same for decimal fractions as for integers. In extracting the square root of decimal fractions, the chief difficulty is encountered in such examples as the following: $\sqrt{.3}$; $\sqrt{.124}$; $\sqrt{1.5}$. The difficulty here is usually due to a mistake in pointing off, or grouping the digits. If a cipher is annexed, so that the number contains an even number of decimal places, this difficulty is removed. For example, $\sqrt{.3} = \sqrt{.30}$; $\sqrt{.124} = \sqrt{.1240}$; $\sqrt{1.5} = \sqrt{1.50}$.

Not all numbers are perfect squares, hence not all numbers can be expressed as the product of two equal factors. For example, if we wish to find the square root of 2, we may annex zeroes to the right of the decimal point and carry the extraction of the root to any required degree of accuracy. $\sqrt{2} = 1.414$; $\sqrt{3} = 1.732$. The square roots of 2 and of 3 are so frequently used in mensuration that they should be memorized.

If we define the square root of an expression to be one of its two *equal* factors it is evident that only abstract numbers can have square roots. No number multiplied by itself equals 9 square feet, or $9, or 9 books. Hence 9 square feet, $9, and 9 books have no square roots.

Geometric Method

The extraction of the square root of 2025 may be explained by the use of a diagram, and this method of explanation may be used in any problem in square root.

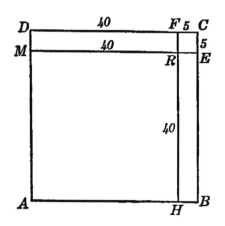

Suppose that ABCD is a square containing 2025 square units. It is required to find the length of one side of this square.

A square whose area is 1600 square units is 40 units on each side, and a square whose area is 2500 square units is 50 units on each side. The side of the square ABCD must therefore contain between 40 and 50 units. If a square whose side is 40 units is taken out, then 1600 square units are taken out and the remaining area is 425 square units. In the figure, AHRM is 40 units on each side, and therefore contains 1600 square units. Since the square ABCD contains 2025 square units, the rectangle MRFD plus rectangle RHBE plus the square FREC contains 425 square inches. The combined length of the two rectangles is known to be 80 units. The problem is to find how wide these rectangles and the square must be so that their combined area shall be 425 square units. If the width were 6 units, the combined area would be more than 6×80 square units, or 480 square units. Try 5 units. The combined area of the rectangles would then be 5×80 square units, or 400 square units, and the area of the square would be 5×5 square units, or 25 square units. The combined area of the rectangles and the square is, therefore, 425 square units. This is known to be the area of the remainder of the square ABCD after the area of the square AHRM has been considered. The length of AB is therefore 40 units plus 5 units, or 45 units.

It should be noticed in the above explanation that the number of units in the combined areas of the two rectangles and the square is the remainder after the number of units in the square AHRM has been subtracted. The sum of the number of units in the length of the two rectangles is the trial divisor, and the number of units in the side of the square RECF is the complete divisor.

The square root of any perfect square or the approximate square root of any number not a perfect square may be found by this method. Whether square root is taught by the algebraic or by the geometric method, the pupils should be asked to formulate a statement of the necessary steps in the process.

Such a statement as the following is valuable as a working rule: 1. Separate the number into periods of two digits each, beginning at the decimal point. 2. Find the greatest square in the left-hand period and subtract it, bringing down the next period. 3. Divide the remainder by twice the part already found. 4. To this divisor add the number thus found and multiply this sum by the number found. 5. Subtract this result and bring down the next period. Proceed as in steps 3 and 4.

Square Root of Fractions

The square root of a common fraction both of whose terms are perfect squares may best be found by extracting the square root of both terms. Thus:

$$\sqrt{\tfrac{4}{9}} = \tfrac{2}{3}: \quad \sqrt{\tfrac{169}{225}} = \tfrac{13}{15}$$

If the terms are not perfect squares, the square root may be found in two ways: (a) by reducing the fraction to a decimal and then extracting the root to the required number of decimal places, or (b) by multiplying both terms of the fraction by a number that will make the denominator a perfect square, and then proceeding as in the illustrations of method "B" below.

Illustrations of Method "A"

$$\tfrac{2}{5} = .4; \sqrt{.4} = .632$$
$$\tfrac{3}{7} = .428571; \sqrt{.428571} = .654$$

Illustrations of Method "B"

$$\sqrt{\tfrac{2}{5}} = \sqrt{\tfrac{10}{25}} = \tfrac{1}{5} \sqrt{10} = \frac{3.162}{5} = .632$$

$$\sqrt{\tfrac{3}{7}} = \sqrt{\tfrac{21}{49}} = \tfrac{1}{7} \sqrt{21} = \frac{4.582}{7} = .654$$

Problems involving the applications of square root occur in the subject of Mensuration.

CHAPTER XX

RATIO AND PROPORTION

One number may be compared with another in two ways; we may inquire how much greater or less one number is than another, or we may inquire how many times one number equals another.

The ratio of two similar quantities is the measure of the relation of the quantities. The ratio of 6 to 3 is 2. The first term of a ratio is called the antecedent; the second term is called the consequent. The ratio expresses how many times the consequent must be taken to produce the antecedent. The ratio of 6 to 3 is 2 because 3 must be taken twice to produce 6.

Proportion arises from a comparison of ratios; it is, therefore, based upon comparison. "It is a comparison of the results of two previous comparisons." Each ratio involves a comparison, and the statement that the two ratios are equal involves a third comparison.

Since a proportion is the expression of equality of two equal ratios, it is an equation.

Proportions were formerly written by use of the symbol ::. Thus

$$4:6::2:3$$

The symbol :: is rapidly disappearing and the sign of equality is being used instead. Thus

(a) $4:6=2:3$

(b) $\frac{4}{6}=\frac{2}{3}$

The equational form (b) is now the one in most common use.

It is easier in practice to place the unknown quantity in the first term.

When proportion is used it should be as a reasoning process and not as a mere mechanical procedure to secure an answer.

The fundamental principle involved in proportion is that the product of the means is equal to the product of the extremes. Because of this principle if any three terms of a proportion are known the fourth may be found. Proportion was formerly called "The Rule of Three," and was regarded as one of the most important parts in arithmetic.

Humpfrey Baker, 1562, speaking of proportion said, "The rule of three is the chiefest and most profitable and the most excellent rule of all arithmetike, for which cause it is said philosophers did name it the golden rule."

The increasing use of the simple equation and of analysis is relegating the subject of ratio and proportion to a position of subordinate importance.

The present tendency to eliminate from the text in arithmetic those problems that do not in some way relate to real life is causing most of the old problems in proportion to be omitted.

The problem which states that 14 men can dig a ditch 8 feet wide, 7 feet deep and 600 feet long in 22 days, working 8 hours a day, and requires the pupil to find out how many days it would require 9 men to dig a ditch 10 feet wide, 8 feet deep and 480 feet long, working 7 hours a day, is omitted from most texts to-day.

Proportion can be used to advantage in some of the applications of arithmetic, especially in the solution of problems relating to similar figures.

CHAPTER XXI

MENSURATION

Why Mensuration is Taught

An examination of the literature treating mensuration shows that the presence of the subject in the elementary schools is justified by widely divergent points of view. Running through the arguments advanced by some of the most distinguished mathematicians is the assumption that the mensuration of the elementary schools should be regarded as a segment of a great system of thought. Naturally, therefore, they urge that it be taught as introductory and preparatory to geometry. Emphasized thus, the purpose of mensuration is to give insight into higher mathematical relations.

Opposed to this point of view is that of some of the more extreme theorists of present day education, who hold that every subject can be justified in the curriculum only on the ground of its immediate utilitarian value. If intimacy between life and the text instead of the exposition of a great system of deductive reasoning is the criterion of the presence of material in the curriculum, then much that has commonly appeared under the title of mensuration in our books should be eliminated.

The wise school administrator and well informed teacher, will see that both aspects of the subject receive attention, but that neither is overemphasized. The great mathematician may be absorbed by the beautiful system of logical reasoning found in geometry, but that is not excuse enough for teaching mensuration as an exemplification of the sys-

tem. A teacher may attempt to articulate the facts and
theories of the school with community activities, as indeed
she should; but that gives her no warrant for neglecting
the preparatory character of mensuration.

Mensuration affords every teacher a great opportunity
of opening the way to new and inviting fields. It is just-
tified partly because it does point the way to a field of
more elaborate constructive thought, partly because it is
immediately useful, but more especially because a knowledge of its materials aids one to properly interpret the
natural features of the world. Instruction in it should
aid in securing an organic conception of certain physical features of the world in which we live. Dominated by
such an aim it would be impossible to teach the subject
without supplying an abundance of valuable information
about computing the contents of solids or the areas
of surfaces. It is through a mastery of such facts that
there gradually dawns upon the pupil a knowledge of the
integral character of the natural world. Mensuration is
not taught primarily to give facts and skill unrelated to
life, but to assist pupils to interpret rationally the universe
about them.

Method to be Employed

This raises the important question as to the method that
shall be employed in teaching the subject. Shall its truths
be demonstrated, illustrated, or taught by rote or formula?
Throughout this entire book we have argued against teaching by formula. We believe, however, that mensuration
suffers more than any other topic in arithmetic in this
particular, for the reason that a large percentage of the
teaching population have no adequate notion of its significance. When teachers as a class are better qualified aca-
demically and professionally for the work they profess to

do, mensuration will be taught on a higher conscious level than now. It will no longer be a habit subject to be acquired through drill, but a thought subject to be developed.

Certainly no comprehensive notion of the interpretative value of mensuration is possible unless its truths are both illustrated and demonstrated. By the time the subject is introduced in the seventh or eighth grade, the children already have a fairly liberal notion of its elementary phases. They learned in the lower grades that mensuration deals with lines, surfaces, and volumes. It remains to apply the ideas thus acquired in the earlier grades to objects of greater utility, such as fields, cisterns, cellars, and the like. It is very necessary that the measurements involved in these larger figures be clearly understood; otherwise both pupils and teachers will continue to make the wildest guesses about distances, areas and contents.

Measurement of Plane Figures

The fact that of all plane figures the rectangle is the only one whose area may be directly found by applying the unit of measure (a square) and then counting the number of times it is applied, should be developed and emphasized. The rectangle is the plane figure from which the mensuration of all other polygons is developed. The area of rectangles, having been studied in the earlier grades, should be profitably reviewed at this point as a basis for the development of the mensuration of other polygons. Attention should be directed to the fact that a square one unit long and one unit wide contains one square unit, and that a rectangle two units long and one wide contains two square units or two times one square unit, which is two square units. Many teachers still permit their pupils to say the area of a rectangle 2 inches long and 1 inch wide is 2 inches × 1 inch = 2 square inches. The area of a rectangle whose length is 4 feet and whose altitude is 3 feet should be expressed as $4 \times 3 \times 1$ sq. ft., or as 4×3 sq. ft. Such a statement gives the correct result, and does not violate any of the principles of multiplication.

Enough class exercises should be given to insure a clear notion of area. Pupils may be required to estimate the number of square inches in their book cover, a window, a picture, a blackboard, and then to test the accuracy of each estimate. The ability to estimate area in square inches, square feet and square yards may be trained by cutting, drawing and by estimating and verifying; of square rods and acres by actual measurement; of a square mile or section of land, by viewing it or walking around it. Later actual practical problems involving area should be given. Perhaps the most practical of these deal with land measure and surveying, flooring, carpentering, lathing, plastering and papering.

The Parallelogram

After the pupil is familiar with the mensuration of the rectangle he should be taught that the area of . any parallelogram is equivalent to that of a rectangle of the same base and altitude as the parallelogram. This may be deduced by a comparison of the

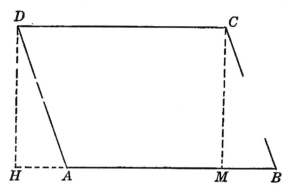

figures. If we draw a perpendicular from C to AB and then cut out the triangle CMB and put it in the position of the triangle ADH, we shall have the rectangle DHMC. This rectangle is evidently composed of the same parts as the original parallelogram ABCD, and is, therefore, equal to it in area. Since the rectangle and parallelogram have bases of equal length and also have equal altitudes the area of the parallelogram may be stated in terms of the rectangle. It should be noted, however, that the area of a general parallelogram cannot be found by applying the square unit of measure, because the square unit will not exactly fit at the vertices. At this stage of the work, the pupil can only infer that the figure is the equivalent of a rectangle because it appears to be so. A rough proof of the equivalence of an oblique parallelogram and rectangle can be secured by folding and cutting off one right triangle and transposing it to the opposite end of the parallelogram. On the other hand the teacher should not overlook the opportunity to tell the pupil that when he learns geometry he will be able to prove conclusively that HMCD is a rectangle.

Measurement of Triangles

The manner in which the various kinds of triangles are related in area to parallelograms and rectangles, should be developed. Before this is done, perhaps, the attention of pupils should be called to the classification of triangles on the basis of the comparative length of their sides or the size of the angles. The essential differences in triangles may be discovered by the children themselves, but the names of the various classes or kinds of triangles should be told the children. An equilateral triangle has all of its three sides equal. An isosceles triangle has two of its sides equal. A scalene triangle has no two sides equal. Triangles are also called acute-angled, obtuse-angled, or right-angled, the classification depending upon the size of the largest angle.

In order to emphasize and fix these conceptions such questions as the following may be asked: Is an isosceles triangle necessarily equilateral? May an isosceles triangle be equilateral? Which is the more general term, equilateral or isosceles? May a right triangle be isosceles? May it be equilateral? May an obtuse-angled triangle be equilateral or isosceles?

The simplest case showing the relation of a triangle to the rectangle is the half square. It seems evident to a child that a diagonal divides the square into halves. This fact can be illustrated by the folding of paper.

The equilateral triangle ABC may be cut up as indicated in the figure, the triangle CDQ being put in the position ARQ and the triangle CDX in the position XMB and the result is the rectangle ABMR.

It can be also shown that an equilateral triangle has half the area of a parallelogram having the same base and altitude as the triangle. By numerous illus-

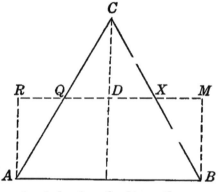

trations similar to these the principle for finding the area of a triangle may be developed.

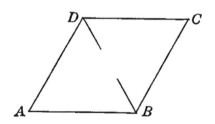

Measurement of Trapezoid and Trapezium

The method of finding the area of a trapezoid may be developed in either of two ways: The points O and E are the midpoints on the lines DA and CB between the parallels DC and AB.

By cutting off the triangle BHE and placing it in the position CME, and placing the triangle AOK in the posi-

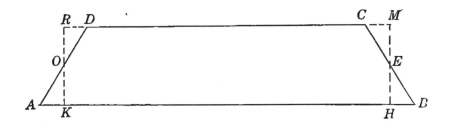

tion ROD, the result is the rectangle KHMR. The altitude of the trapezoid has not been altered and the sum of the bases of the original trapezoid ABCD is equal to the sum of the bases of the resulting rectangle.

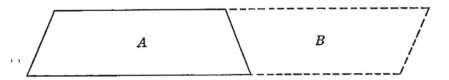

The same result may be derived by constructing the trapezoid B equal to the trapezoid A. The result is a parallelogram. The altitude of the parallelogram is the same as that of the trapezoid, and half of the sum of the bases of the original trapezoid is equal to one base of the parallelogram. The area of the trapezoid can now be stated, since the area of the parallelogram is known.

Measurement of Circles

Pupils should be required to learn three formulas for finding the area of a circle after the method commonly found in most good texts has somewhat justified the process to the pupil's mind. Since the number of units in the area of a circle is equal to one-half the number of units in the circumference times the number of units in the radius, we have the formula $a = \dfrac{C\,r}{2}$, where a, c, and r represent the number of units in area, circumference, and radius respectively. From this formula the other two formulas for the area of the circle may be easily deduced. If the teacher will have her pupils carefully measure the circumferences and diameters of several circles of different sizes (a dollar, tin cup, or any circular object whose dimensions can readily be found will serve the purpose), and then have the pupils divide the length of the circumference by that of the diameter, the quotient in every case will be found to be about $3\frac{1}{7}$. This ratio of the length of the circumference to that of the diameter is proved in geometry to be always the same, and to be approximately equal to 3.14159. For all practical school purposes the values 3.1416 or $\frac{2\,2}{7}$ are sufficiently accurate.

Since the ratio of the length of the circumference to that of the diameter is constant, a special symbol is used to represent this value. This symbol is the Greek letter *pi*. We therefore say $\dfrac{C}{D} = \pi$, or $\dfrac{C}{2r} = \pi$, since $D = 2r$, or $C = 2\pi r$. If we substitute for C its value $2\pi r$ in the formula $a = \dfrac{C\,r}{2}$ we shall have $a = \dfrac{2\pi\,r \times r}{2}$, or πr^2. Since $d = 2r$, therefore $r = \dfrac{d}{2}$

and $r^2 = \dfrac{d^2}{4}$. If we put this value for r^2 in the formula πr^2, we shall have $a = \pi \dfrac{d^2}{4}$.

Since $\pi = 3.1416$, we have $a = \dfrac{3.1416d^2}{4}$, or $.7854d^2$.

The pupil should know these three formulas for area:

$$\frac{C\ r}{2}, \quad \pi\ r^2 \text{ and } \pi\ \frac{d^2}{4}$$

Measurement of Rectangular Solids

Just as the area of most plane figures may be found by comparing them with the area of a rectangle, so we may find the volume of regular solids by comparing them directly or indirectly with the volume of the rectangular solid. The rectangle is the only plane figure whose area may be determined by applying the unit of measure, a square, and then counting the number of times it is contained in the rectangle whose area is desired, and the rectangular solid is the only solid whose volume can be found by applying the unit of measure, a cube, and counting the number of times it is contained in the solid whose volume is required.

If the teacher has a few small cubes and will build up before the class, or have some pupil build up, rectangular solids of various dimensions the fact that the number of units in the volume of a rectangular solid is equal to the product of the number of units in its three dimensions can be easily deduced. Thus, the volume of a rectangular solid whose dimensions are 4 in. by 3 in. by 2 in. is easily seen to be $4 \times 3 \times 2 \times 1$ cu. in. $= 24$ cu. in. It is very desirable that the teacher should not permit the statement so commonly used in such problems (4 in. $\times 3$ in. $\times 2$ in. $= 24$ cu. inches). (See chapter on Accuracy.)

The Use of Models

Space does not permit an extended discussion of the various methods whereby the volumes of the various solids, commonly considered in the arithmetic of the grades, may be shown to depend directly or indirectly upon the volume of the rectangular solid. Most arithmetics contain a sufficiently detailed discussion of these facts. The best interests of the class will frequently be served if the pupils have access to models of the various solids that are to be studied. Inexpensive models for this purpose can usually be purchased at a comparatively small expense. Many of the more common solids can be easily made out of wood or pasteboard by the teacher or the pupils. In studying the pyramid, the cylinder, the cone, and the sphere it is very desirable to have models at hand, as they will frequently make clear some difficulty due to the inability of the pupil to properly image the solid under consideration. It is undoubtedly true that such models may be used to excess and thus defeat the very purpose that they are to serve, i. e., to enable the pupil to image clearly the form that is under consideration. That models may be used to excess in the study of mensuration should not condemn the use of models, but should argue for more wisdom in their use. It is wise to dispense with the use of objects in mensuration just as soon as the pupil can properly image the figure under discussion. Some pupils can image much more readily and more distinctly than others, and the teacher should be careful to use the objective material only in those cases where it seems imperative.

Measurement of Cylinder

It is desirable in studying the lateral area of a cylinder to have the pupil think of a piece of paper so cut that it will just cover the lateral surface when wrapped about it. When the paper is removed the pupil will readily see that it is in the form of a rectangle. The base of the rectangle is equal in length to the circumference of the cylinder and the altitude of the rectangle is equal to the altitude of the cylinder. Since the number of units in the area of the rectangle can be found, the number of units in the lateral surface of the cylinder can readily be determined. Let the pupil try to image the shape of the paper that could be made to just cover the lateral surface of a right cone. He should see that the paper is in the form of a sector of a circle. From this the formula for the lateral surface of a cone may be determined.

Measurement of Pyramids and Prisms

For comparing the volume or cubical contents of a right pyramid with that of a right prism of equal base and altitude it is well to fill the pyramid with some substance such as sand and pour it into the prism. When the pupil finds that the pyramid must be filled three times and the contents poured into the prism in order that the latter may be full, he is ready to infer, as is proved in geometry, that the volume of a pyramid is equal to one-third of the volume of a right prism of equal base and altitude. The volume of a cone may be compared with that of a cylinder of equal base and altitude by the same method.

Measurement of the Sphere

Pupils are usually interested in the experiment to determine the area of the surface of a sphere. Cut a wooden ball through the center by a plane. Place a tack at the center of the sphere and let a pupil wind tape about this sufficient to just cover the surface of the great circle. Let the pupil wind sufficient tape about the hemisphere to just cover its curved surface and compare the amount of tape required to cover the two areas. If the work has been carefully done, it will be found that twice as much tape is required to cover the hemisphere as to cover the great circle of the sphere The number of units in the area of the great circle of the sphere is know to be πr^2, where "r" represents the number of units in the radius,—therefore, the number of units in the curved surface of the hemisphere is $2\pi r^2$ and the surface of the entire sphere is $4\pi r^2$.

Use of Literal Representation

Most pupils see the advantage of letting "s" represent the number of units in the side and "a" the number of units in the altitude. If the pupils are required to use these and other abbreviations in the work of the eighth grade a large saving of time may result.

Problems

It is now proposed to consider in some detail a few problems somewhat more difficult than most of those in the arithmetic of the grades. Such a consideration should be valuable from the teacher's standpoint, because it may extend the margin of scholarship somewhat and because of methods that are suggested for the solution of such problems. The problems that follow are of a type that are a good test of the ability of the pupil to image clearly.

1. Compare the area of a square with that of the largest circle that may be cut from it.

In solving such a problem it is necessary that the pupil should image the figure clearly and should see what dimensions the circle and the square have in common. If the pupil is not able to image this relationship readily the figure may be drawn. It is seen that the side of the square is equal in length to the diameter of the circle. Here then is a basis for comparing their areas.

Let $r =$ the number of units in the radius of the circle.

Then $2r =$ the number of units in the diameter of the circle.

Therefore, $2r =$ the number of units in the side of the square.

The area of the circle is πr^2.

The area of the square is $(2r)^2$ or $4r^2$

Therefore the area of the circle is to the area of the square as πr^2 is to $4r^2$ or $\dfrac{\pi r^2}{4r^2} = \dfrac{\pi}{4}$.

Since $\pi = \frac{22}{7}$ we have $\dfrac{\left(\frac{22}{7}\right)}{4} = \frac{22}{28}$ or $\frac{11}{14}$.

Therefore the area of the circle is $\frac{11}{14}$ of the area of the square. This result is true irrespective of how large or how small the given square may be.

2. The problem to compare the area of a circle with the largest possible square that can be cut from it, is slightly more difficult. In this problem the diameter of the circle is seen to be equal to the diagonal of the square.

Let $r =$ the number of units in the radius of the circle.

Let $2r =$ the number of units in the diagonal of the square.

Since the number of units in the diagonal of the square $= 2r$, the number of units in the square of one side is easily found to be $2r^2$ This is the area of the square.

The area of the circle is πr^2.

The area of the circle is to the area of the square as

$$\dfrac{\pi r^2}{2r^2} = \dfrac{\left(\frac{22}{7}\right)}{2} = \frac{22}{14} = 1\frac{1}{7}.$$

The fact that such problems are considered also in geometry should not deter the teacher from considering them in the more advanced work in mensuration in the grades.

3. Compare the volume of a cube with the volume of the largest possible sphere that can be cut from the cube. It is easily seen that the length of the diameter of the sphere is equal to the length of the edge of the cube.

Let r = the number of units in the radius of the sphere.

Therefore $\dfrac{4\pi r^3}{3}$ = the number of units in the volume of the sphere.

$2r$ = the number of units in the edge of the cube.

Therefore $8r^3$ = the number of units in the volume of the cube.

The ratio of the volume of the cube to the volume of the sphere is therefore,

$$\frac{8r^3}{\frac{4\pi r^3}{3}} = \frac{6}{\pi} = \frac{6}{\frac{22}{7}} = \tfrac{42}{22} = \tfrac{21}{11}.$$

The volume of the cube is, therefore $\tfrac{21}{11}$ of the volume of the sphere. It would be an easy problem to compare the surface of the cube with that of the sphere.

It is sometimes well to put the best pupils of the class on their mettle and such problems are well adapted to this purpose.

Observational Geometry

There is a tendency to curtail the work in mensuration, which formerly consisted largely of definitions, formulas and problems, and to include some elementary geometry in the course of the seventh and eighth grades. The type of work introduced is variously known as constructional, observational, inventional, intuitional, or concrete geometry. Its purpose is to acquaint the pupil with the more important geometrical concepts, to train the eye and the hand in the use of the straight edge, compasses, triangle and protractor, and to develop the powers of observation and intuition as applied to geometrical forms.

It is desirable that pupils who do not enter high school should have some knowledge of the fundamental concepts of geometry. An introductory course in geometry prepares those pupils who enter high school for the more formal study of the subject.

Some work in introductory geometry is now given in about 25% of the larger schools of the country. The instruction is sometimes made an integral part of the course in arithmetic and is included under the topic of mensuration; in some schools one or more periods a week are regularly devoted to the work in the seventh and eighth grades and the connection with mensuration is not so immediate.

In many of the European schools the teacher of mathematics is also the teacher of drawing, and work of the kind referred to above is made a part of the instruction in drawing. It has been suggested that one reason for the slowness of the introduction of this type of work into American schools is that some work in "mechanical" drawing is usually given in the art courses. When such has been the case emphasis has usually been placed upon artistic results rather than upon the training of the power of observation and generalization, because our art teachers do not usually emphasize the subject from the point of view of mathematics. The schools in which these constructional exercises are taught usually attempt to correlate the work as closely as seems practicable with the course in manual training.[1]

In several of the German states the course in arithmetic includes the study of some of the simple geometric forms

[1] See "Mathematics in Elem. Sch. of U. S." Report to International Commission, p. 136.

and constructions as early as the fifth and sixth years. The pupils become familiar with the fundamental properties of straight lines, angles, triangles, quadrilaterals, polygons, and circles, and models of some of the simpler geometrical solids are constructed.

Simple and easily available material for the study of elementary geometrical forms is abundant, and the course in mathematics in the grades may be both enriched and vitalized by a judicious use of this material. Many of the most important theorems of geometry may be intuitively established by the use of constructions and measurement. A better insight is also gotten into most of the rules and formulas of mensuration.

Pupils of the grammar grades should be taught the meaning of such terms as perpendicular, right angle, parallel, etc. They should be required to construct at the blackboard and by the use of paper or cardboard the right, isosceles, equilateral and scalene triangles and by means of cutting out, tracing, applying and the use of the protractor they should discover experimentally the elementary truths about congruency of triangles and parallelograms, and the size and relation of angles in the various triangles. The formulas for the area of the triangle, parallelogram and trapezoid should be discovered through drawing, paper cutting or by means of tracing. In some schools plotting paper is extensively used as a means of estimating areas.

Congruent and Similar Figures

The properties of congruent and of similar figures may be used by the pupils in a variety of interesting and instructive ways. Ask the pupils to devise two or three ways for determining the height of a tree or a building on the school grounds. The following methods are suggested: (a)

By comparing the length of the shadow of a tree or building with that of a pole of known height. (b) By use of an isosceles right triangle. If AB = BE, then AC = CD, therefore to find CD measure AC. (c) By drawing to scale.

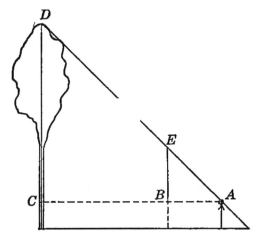

Let pupils determine the distance between two inaccessible points by use of congruent triangles.

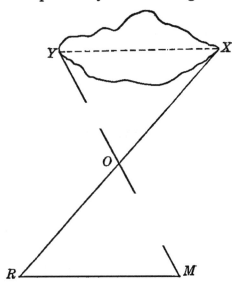

Set up a pole at O and measure the distances OX and OY. On 'a line with X and O put a stake at R such that the distance OX = OR. On a line with Y and O put a stake at M such that OY = OM. Measure the distance RM. This will be equal to XY, which is the required distance.

Ask pupils to devise methods of determining the width of a river without crossing it.

The following methods may be used:

First Method. To determine the distance A B.

A pupil who has an isosceles right triangle may walk along the bank sighting along the hypotenuse A C and

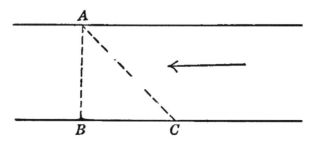

holding one arm parallel to the line B C, until he reaches a point, C such that the point A on the opposite bank from B is seen. The distance B C is then equal to A B.

Second Method. To determine the length of A B.

Mark off a line B M perpendicular to A B and place stakes at points B and M. Erect perpendicular M R of a

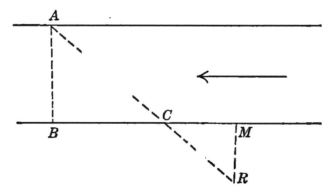

convenient length to line M B. Place the stake at C where the line B M intersects the imaginary line A R. The distance A B is then found from the proportion $\dfrac{A B}{M R} = \dfrac{B C}{M C}$, in which M R, B C and M C are known.

CHAPTER XXII

GRAPHS

Newspapers, popular magazines and trade journals make frequent use of graphs in order to make clear the relations between magnitudes. It is desirable that some instruction in this subject should be given in the schools. The pupil of the seventh or eighth grade is more or less familiar with the practice of representing the relative sizes of armies and navies by men of various heights. He has seen lines of various lengths used to represent the growth of the population and he knows that curves are sometimes used to represent variations in temperature or fluctuations in the prices of certain commodities. In all of the more progressive European countries the subject of graphs is systematically taught before the pupil enters what corresponds to our secondary school. Teachers in this country are beginning to appreciate the fact that the graph may be used to advantage to illuminate certain topics and to emphasize the relations between various magnitudes. The graph is an effective method of impressing upon the pupil relations that might otherwise be obscure. The underlying principles are easily understood, and some consideration should be given to the subject in the sixth, seventh, or eighth grades. Rates of increase and decrease are always shown more clearly by the use of graphs than by tables of statistics. When two or more graphs are drawn on the same scale and with the same axes, comparison of the variations can be readily made. The graph should not be taught as an end in itself, but as a means to an end. It should be used to supplement, not to supplant analyses.

The graph is a natural place for the introduction of the *function* idea which is extensively used in higher mathematies. Much more emphasis is placed upon the idea of function in the elementary and secondary schools of Europe than in this country. It is probable that within a few years we shall place more emphasis upon this idea in our schools. When magnitudes are so related that any variation of one causes a variation of the other, each is said to· be a function of the other. For example, the distance that a train will run in a certain time is a function of the rate, because the distance varies as the rate varies. The interest at a given rate is a function of the principal and of the time, because as these vary the interest varies. The area of a triangle is a function of its base and its altitude, and the area of a circle is a function of its radius.

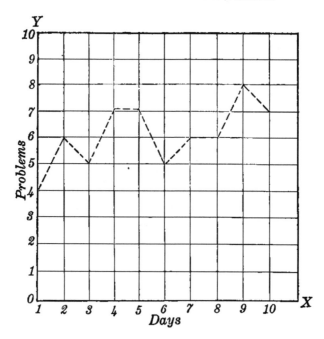

The topic may be introduced by showing the pupil how he may represent his daily record in a given subject.

The preceding graph represents the achievement of a pupil in solving some problems in arithmetic in a limited time. The graph shows the number solved correctly for ten consecutive days. On the horizontal axis "OX" the days are represented, and on the vertical axis "OY" the number of problems solved correctly is shown. The graph shows that on the second day six problems were solved; on the third day, five; on the ninth day, eight:

Most pupils like to graph the daily variations in temperature. Interesting graphs may be made showing the fluctuations in temperature at a given hour—for example, 9 A. M.—for several consecutive days. The graph below illustrates this.

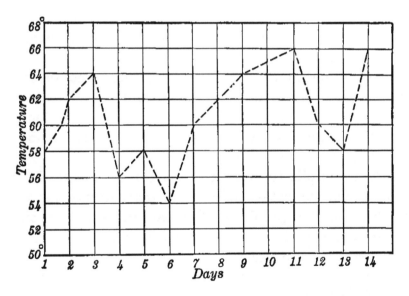

At 9 A. M. on the first day the temperature was 58°; on the sixth day it was 54°; on the 11th day it was 66°.

Ask the pupils to graph the growth of the population of

the city or county in which they live, and then, using the same axes and the same scale, graph the growth in school population for the city or county. Pupils will be interested in comparing these graphs. Similar graphs may be made for the state or the nation. Statistics for such graphs may be readily secured in most communities.

Numerous interesting facts whose relations are emphasized by the use of graphs may be easily found by teacher and pupil. Most pupils from rural communities will be interested to graph the yearly yield of wheat, corn, oats, barley, potatoes.

County, state, and government reports furnish numerous statistics that are of interest to pupils.

If the school engages in athletic contests, some of the pupils who are interested in these activities may be asked to graph the results of a series of contests. The number of points or of goals scored by each player or each team may be graphed.

Some pupils in the school may be interested to keep a graphic record of the attendance and to compare this graph with the graph of various facts that affect the attendance, such as temperature, rainy days, number of cases of sickness in the community, etc.

Ask the pupils to make a graph for the simple interest on $1 at various rates, and require them to use this graph in solving several problems. Compare the graph for simple and for compound interest on a given principal, at a given rate, for 6, 7, or 8 years.

Show the pupils how the graph may be used in the solution of problems. For example: A and B start from the same place and travel in the same direction. A travels at the rate of 4 miles an hour and B at the rate of 6 miles an hour. If A has three hours start, in how many hours will B overtake A?

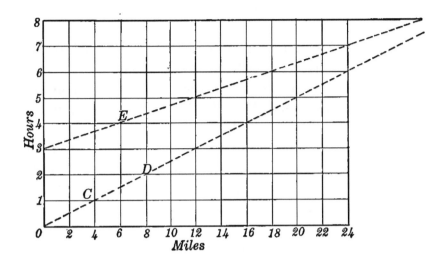

The graph shows that at the end of the first hour A is at C and his path is OC. At the end of the second hour he is at D. One hour after B starts he is at E. B overtakes A when his path crosses that of A. Determine by use of a graph how many hours this is after B starts, and how far each has traveled.

Request the pupils to graph some facts in which they are interested, and from these graphs ask other pupils to state some of the facts.

NOTE.—Teachers who wish an elementary book on the subject of graphs are referred to Auerbach's ''An Elementary Course in Graphic Mathematics.'' Published by Allyn & Bacon.

CHAPTER XXIII

SHORT CUTS

The "*why*" is of importance in the teaching of arithmetic, especially after the fifth or sixth school years. In the later years of the grammar grades the "*how*" should receive no less emphasis than in the earlier year, but the "*why*" should receive gradually increasing emphasis. To many pupils arithmetic is composed of numerous, dogmatically stated rules and scores of definitions; they have derived but little from that richer part of arithmetic which emphasizes the thought side.

When Short Cuts Should be Introduced

However, in the upper grammar grades there are times when "*the how*"—the mere doing of a thing in the correct way—should receive no little emphasis. After the fundamental operations have been mastered and the general underlying principles of arithmetic are understood, certain short cuts may be introduced with profit.

It is not intended that the teaching of short cuts should be postponed until the arithmetic of the grades has been practically completed, but it is best that before a pupil is taught a short cut he should know the longer method of securing the same result. The longer method is usually more easily explained and is more likely to appeal to the understanding of the pupil. If arithmetic were taught merely as a tool to be used in order to obtain correct results in computation, the short cut should be introduced earlier than is here advocated.

How to Teach Short Cuts

Short cuts, if properly presented, will engender much interest and enthusiasm in the subject and will save time in future computations. In teaching short cuts the teacher should carefully emphasize the fact that there is not a single short process in grammar school arithmetic that cannot be explained. Some of them are easily proved by the use of elementary algebra. It is neither necessary nor desirable that many of these proofs should be presented to a class, but the confidence of the pupil will be increased, if he is assured that a given short cut can be readily explained and that it is not due to the so-called ''mystery of numbers.''

A short cut becomes really valuable to a pupil after he has such a mastery of it that he recognizes it under whatever conditions it may be presented. One that has not been thoroughly mastered and impressed upon the mind by frequent drill is of comparatively little value. When an occasion arises in which the short cut may be used to advantage, the mind trained to use the longer method fails to utilize the short cut. A few of these short processes thoroughly mastered are of more value than many of them not well enough known to be used upon the proper occasion. Any short cut that is learned mechanically will be forgotten within a short time after the drill upon it ceases, unless the pupil has such a thorough mastery of it that he uses it almost as mechanically as he should use the facts of the multiplication table, and unless his work is of such a nature that frequent opportunity to use the short cut is presented. Choose a few of the short cuts which seem especially valuable, master them thoroughly and be continually upon the lookout for an opportunity to use them. When a few of them have been so mastered that their application has become more or less mechanical a few more should be taken up in the same way. The importance of short cuts can be easily over-emphasized. It would be unwise for any teacher to attempt to have the class master all the short processes that are suggested here.

Short Cuts in Multiplication

Multiplication is a short cut for addition. When it is required to multiply 47 by 18 the question really is, "what is the result when 47 is used 18 times as an addend?" Our mastery of the multiplication table and our knowledge of place value enables us to obtain the required result by a shorter process than addition. The example just cited could not be so quickly worked by one who is not familiar with the multiplication table unless he used some mechanical device to aid him in the computation The Russian peasant who can add and can multiply and divide by 2 could work the above example, but he could not work it in the way most familiar to us. He would proceed as follows:

47 ×	18
23	36
11	72
5	144
2	288 (omit)
1	576

Each number in the first column is divided by 2; if the quotient is not an integer only the integral part of the quotient is·taken; this division is continued until the quotient 1 is reached. In the second column each number is multiplied by 2; this is continued until as many numbers have been obtained as in the first column. All numbers in the second column, except those which stand opposite even numbers in the first column are added. The sum thus obtained is the required product; i. e., 576+144+72+36+18 =846, or 47×18. It is apparent that the term "short cut" is a relative term. If our method of multiplying could be taught to a Russian peasant it would be a valuable short cut for him.

Professional mathematicians use numerous short cuts for obtaining results. Some of these, such as tables of logarithms; tables of squares and cubes, and of square roots and cube roots, save an enormous amount of time in computation, but not all of these can be used to advantage by a pupil in the upper grammar grades.

Attention will be directed first to some short cuts that can be used to advantage in multiplication.

To square any number ending in 5. Multiply the number of tens by one more than itself and annex 25. For example, to square 35. The number of tens is 3. Multiply this by 1 more than itself and the product is 12. Annex 25 and the final result is 1225. By the same rule the square of 45 is found to 2025; and the square of 75 is 5625. $105^2 = 11025$.

The above is obviously applicable, with modifications, to examples like the following: $(7\frac{1}{2})^2$, $(9\frac{1}{2})^2$, $(4\frac{1}{2})^2$. But it is easier to work such examples by the following rule: Multiply the integer by one more than itself and add $\frac{1}{4}$. Thus, $(7\frac{1}{2})^2 = (8 \times 7) + \frac{1}{4} = 56\frac{1}{4}$. $(9\frac{1}{2})^2 = (10 \times 9) + \frac{1}{4} = 90\frac{1}{4}$. $(4\frac{1}{2})^2 = (5 \times 4) + \frac{1}{4} = 20\frac{1}{4}$. It is evident that the $\frac{1}{4}$ may be written as .25.

Required to multiply two numbers whose ten's digits are the same and whose unit's digits add to make 10. For example, 47×43, or 84×86. Multiply the ten's digit by one more than itself and annex the product of the unit's digits. Thus, 47×43 equal 2021. 84×86 equal 7224. 13×17 equal 221. Whenever the product of the unit's digits is a one digit number, a zero must be put in the tens' place. For example, if 81 is to be multiplied by 89, the product of the unit's digits (1 and 9) is a one digit number, hence the final result is 7209. Similarly the product of 51 and 59 equal 3009. The rule for squaring a number ending in 5 is evidently a special case of the rule just cited.

To multiply two numbers between 10 and 20. For example, 17×15, or 14×17, or 19×16.

To either of the numbers add the unit's digit of the other. Annex a cipher to this result and add the product of the unit's digits. Example, 17×15. $17 + 5$ equal 22 (or $15 + 7$ equal 22). Annex a cipher and the result is 220. To this add the product of the unit's digits and the result is $220 + 35$, or 255.

Example, 14×17. $14 + 7 = 21$. Annex a cipher and the result is 210. Add 4×7 to this, and the result is 238.

This computation may always be made orally with ease and is a valuable short cut.

When one factor is as much greater than some multiple of 10 as the other is less than that same multiple.

Example, 28×32. (32 is 2 more than 30 and 28 is 2 less than 30).

Example, 46×54. (54 is 4 more than 50, and 46 is 4 less than 50).

Square the multiple of ten and from this result subtract the square of the difference between one of the given numbers and the multiple of 10.

Example, 28×32. 30 is the multiple of 10 between these numbers. $30^2 = 900$. The difference between one of the given numbers and 30 is 2. $2^2 = 4$. $900 - 4 = 896$, which is the required product.

Example, 46×54. $50^2 = 2500$. $4^2 = 16$. $2500 - 16 = 2484$.

Example, 39×41. $40^2 = 1600$. $1^2 = 1$. $1600 - 1 = 1599$.

Those who understand the elements of algebra will recognize the above short cut as $(a - b)\,(a + b) = a^2 - b^2$, since we may express 28×32 as $(30 - 2)\,(30 + 2) = 30^2 - 2^2 = 900 - 4$ equal 896.

The product of any 2 numbers equals the square of the number midway between them, minus the square of half their difference.

This comes under the same algebraic law as the short cut preceding.

Example, $17 \times 13 = 15^2 - 4 = 221$.

Find the cost of 19 books at 13c each. Result in cents $= 16^2 - 9 = 247$.

By the complement of a number is meant that number which added to the given number makes the next higher power of 10. Thus the complement of 98 is 2; of 87 it is 13, of 992 it is 8.

To multiply together two numbers whose complements are not large and whose complements are computed from the same power of ten, proceed as follows:

Example. 98×95.   98   2 = complement of 98.

                        95   5 = complement of 95.

98 − 5 and then annex 10, (5×2) $\overline{93\ 10}$ = required product.

From either number (98 or 95) subtract the complement of the other number. To this result annex the product of the complements.

Example. 89×93.   89   11 = complement of 89.

                        93   7 = complement of 93.

89 − 7 and then annex 77, (7×11) $\overline{8277}$ = required result.

Example. 989 × 988.   989   11 = complement of 989.

                        988   12 = complement of 988.

Subtract 12 from 989 and $\overline{977132}$ = required result.
annex 132, (11×12).

When the complement is computed from 100 and the product of the complements is not a two digit number a zero must be put in tens' place. When the complement is computed from 1000 and the product of the complements is not a three digit number, a zero must be put in hundreds' place, or in hundreds' and and in tens' places.

Example, 97×99. The product of the complements (3 and 1) is not a number of two digits. The result is 9603.

Example, $998 \times 994 = 992012$ (2×6 is only a two digit number).

Example, $999 \times 997 = 996003$ (1×3 is only a one digit number).

By the supplement of a number is meant that which subtracted from the number makes the next lower power of ten. Thus the supplement of 104 is 4; the supplement of 111 is 11; the supplement of 1007 is 7.

To multiply together two numbers whose supplements are not large and whose supplements are computed from the same power of ten, proceed as indicated below:

Example, 112×103. 　　　112　12 = supplement of 112.
　　　　　　　　　　　　　　103　 3 = supplement of 103.

$112 + 3$, then annex 36, (3×12) $\overline{11536}$ = required product.

To either number, 112 or 103, add the supplement of the other. To this result annex the product of the supplements.

Example. 117×105. 　　　117　17 = supplement of 117.
　　　　　　　　　　　　　　105　 5 = supplement of 105.

$117 + 5$, then annex 85, (5×17) $\overline{12285}$ = required product.

When the product of the supplements does not give a two digit number a zero must be put in tens' place if the supplement is computed from 100. If the supplement is computed from 1000 and the product of the supplements is not a three digit number, a zero must be put in the hundreds, or the hundreds' and the tens' places.

Example, 102×104. The product of the supplements (2 and 4) is a one digit number. The result is 10608.

Example, $1014 \times 1003 = 1017042$.

After a short drill the short cuts involving complements and supplements can be quickly applied without the complements or supplements being written down.

The Elevens Rule

Example, 11×4532.

Write 2 for the right hand figure of the product. Add the 2 and 3 for the next figure of the product. Add 3 and 5 for the third figure of the product; add the 5 and 4 for the fourth figure and write 4 for the fifth figure of the product. The product is, therefore, 49852. The reason for this procedure is easily seen by considering the longer form and seeing the additions that must be made.

$$
\begin{array}{r}
4532 \\
11 \\
\hline
4532 \\
4532 \\
\hline
49852
\end{array}
$$

Similarly $11 \times 482 = 5302$.

Aliquot Parts. By the aliquot parts of a number are meant those numbers which are contained in the given number without a remainder. Thus 4 is an aliquot part of 12; $2\frac{1}{2}$ is an aliquot part of 5. When aliquot parts are used in arithmetic the given number is usually ten, a hundred, or a thousand.

Aliquot parts of a number may frequently be used to advantage in multiplication.

Example. To multiply 84 by 25 we may annex two ciphers to the 84 (that is, multiply it by 100) and then divide the result by 4, since $25 = \frac{1}{4}$ of 100.

Similarly to multiply a number by $16\frac{2}{3}$ we may first multiply the number by 100 and divide the result by 6.

To multiply a number by $2\frac{1}{2}$ we may multiply the given number by 10 and divide this result by 4, since $2\frac{1}{2}$ equal $\frac{1}{4}$ of 10.

The pupil should learn the commonly used aliquot parts of 10, 100 and 1000, as they can frequently be used to advantage, as suggested above.

Frequently used aliquot parts of ten

$2\frac{1}{2}$	$=\frac{1}{4}$ of 10
$3\frac{1}{3}$	$=\frac{1}{3}$ of 10
5	$=\frac{1}{2}$ of 10
$6\frac{2}{3}$	$=\frac{2}{3}$ of 10

Frequently used aliquot parts of one hundred

50	$=\frac{1}{2}$ of 100	$12\frac{1}{2}$	$=\frac{1}{8}$ of 100
$33\frac{1}{3}$	$=\frac{1}{3}$ of 100	$11\frac{1}{9}$	$=\frac{1}{9}$ of 100
25	$=\frac{1}{4}$ of 100	10	$=\frac{1}{10}$ of 100
20	$=\frac{1}{5}$ of 100	$9\frac{1}{11}$	$=\frac{1}{11}$ of 100
$16\frac{2}{3}$	$=\frac{1}{6}$ of 100	$8\frac{1}{3}$	$=\frac{1}{12}$ of 100
$14\frac{2}{7}$	$=\frac{1}{7}$ of 100	$6\frac{1}{4}$	$=\frac{1}{16}$ of 100

The equivalents of the following parts of 100 should also be learned, $\frac{2}{3}$, $\frac{3}{4}$, $\frac{2}{5}$, $\frac{3}{5}$, $\frac{5}{6}$, $\frac{2}{7}$, $\frac{3}{7}$, $\frac{5}{8}$, and $\frac{7}{8}$.

The pupil should learn also the more frequently used aliquot parts of 1000, such as $\frac{1}{3}$, $\frac{1}{4}$, $\frac{1}{8}$, and $\frac{2}{3}$.

If required to find the cost of 42 articles at $16\frac{2}{3}$ cents each, the pupil should utilize his knowledge of aliquot parts. At \$1 each the articles would cost \$42, therefore, at \$$\frac{1}{6}$ each, the cost will be $\frac{1}{6}$ of \$42, or \$7.

If the problem requires the cost of 24 articles at $37\frac{1}{2}$ cents each the pupil should be trained to see that the cost will be just $\frac{3}{8}$ of \$24, or \$9.

To multiply two mixed numbers when the fractions are each $\frac{1}{2}$:

To the product of the whole numbers add one-half the sum of the whole numbers and to this add one-fourth.

Example, $4\frac{1}{2} \times 7\frac{1}{2}$. $4 \times 7 = 28$. $\frac{1}{2}$ the sum of the whole numbers $(4+7)$ is $5\frac{1}{2}$. Add this to 28 and to this sum add $\frac{1}{4}$. The final result is $33\frac{3}{4}$.

Similarly, $5\frac{1}{2} \times 9\frac{1}{2}$. $5 \times 9 = 45$. $\frac{1}{2}$ of $(5+9)$ is 7. $45 + 7 + \frac{1}{4} = 52\frac{1}{4}$.

CHAPTER XXIV

LONGITUDE AND TIME

The subject of Longitude and Time is common to both Arithmetic and Mathematical Geography. In recent years the tendency has been to omit the subject from courses in arithmetic, but it is still taught in so many schools that its treatment is necessary here. Longitude and time is a practical subject to the navigator and to the astronomer. The rapid adoption of standard time by most of the civilized nations of the world has rendered impractical the older text-book problems in longitude and time. The old-style, complicated problems of a generation ago are giving way to practical problems involving standard time.

The principles underlying the subject can be easily explained, and no teacher should permit the substitution of arbitrary rules for the application of a few simple principles. Many pupils of the grades obtain but little of value from a study of the subject because it is presented in the form of rules.

In this discussion it is assumed that the pupil knows the meaning of the terms degree, longitude, and meridian. It is assumed also that he is familiar with the units of circular measure.

There is only one equator from which latitude may be reckoned, but longitude may be reckoned from any meridian. Prior to 1844 it was the custom in many countries to use the longitude of their capital as the zero of longitude. Since 1844 the meridian running through the observatory at Greenwich, England, has been most generally used.   Greenwich is about five miles southeast of London.

Pupils will be interested to know the accuracy with which longitude can be determined by a skillful astronomer with refined instruments.   The difference in longitude between two cities on the same continent can be determined within six or eight yards; the difference in longitude between Washington and Greenwich is known within about three hundred feet.

The Tables

The tables for solving the problems of longitude and time may be developed either by considering the real rotation of the earth upon its axis or the apparent revolution of the sun around the earth.

Every point on the earth's surface, except the poles, is carried, by the earth's rotation, through 360° in 24 hours.

From this fact the tables are developed:

Since in 24 hr. the earth turns through 360°,

therefore in 1 hr. it turns through $\frac{1}{24}$ of 360° or 15°,

therefore in 1′ it turns through $\frac{1}{60}$ of 15°, or $\frac{1}{4}$°, or 15′,

therefore in 1″ it turns through $\frac{1}{60}$ of 15′ or $\frac{1}{4}$′, or 15.″

Since the earth turns through 360° in 24 hr.,

We may say:

therefore it turns through 1° in $\frac{1}{360}$ of 24 hr., or $\frac{1}{15}$ hr. or 4 min.,

therefore it turns through 1′ in $\frac{1}{60}$ of 4 min., or $\frac{1}{15}$ min. or 4 sec.,

therefore it turns through 1″ in $\frac{1}{60}$ of 4 sec., or $\frac{1}{15}$ sec.

The mastery of the preceding tables is of importance in a detailed study of longitude and time.   The teacher should not permit the pupil to use the statement found in many text-books, that 15° = 1 hour.   There is no equality between longitude and time in the sense that "equality" is used in mathematics. Two times 4 equals 8 is one of the fundamental facts of mathematics, whereas 15° corresponds to 1 hour of time only because there are 360° in a circumference and because the earth rotates upon its axis once in 24 hours. If the rotation period of the earth were 20 hours instead of 24 hours, 15° would no longer correspond to 1 hour; 18° would correspond to 1 hour. After the facts of the tables have been mastered, numerous easy problems based upon them should be given. The problems below will suggest others to the teacher:

Two cities differ in longitude by 30°, by 45°, by 60°, by 150°, by 22°.30', by 7°; what is the difference in time?

The difference in time between two cities is 3 hours, 5 hours, 1½ hours, 3⅓ hours; what is the difference in longitude?

To Determine Difference in Longitude

Pupils are sometimes uncertain whether the longitude of two given places should be added or subtracted in order to obtain the difference in longitude. This uncertainty may be eliminated by illustrations similar to those below. The teacher should give the illustrations a local setting by substituting for A, B, and C objects that are familiar to the pupil. The distance from the schoolhouse to certain places in opposite directions from it, or distances from the city in which the school is located to two other places in opposite directions from the home city, may be used.

B   20 mi.   A   30 mi.   C   In this illustration the cities A, B, and C are in the same straight line, A being 20 miles from B and 30 miles from C. What is the distance from B to C? The pupil will readily see that the distance from B to C is 20 miles plus 30 miles, or 50 miles. If the distance from B to C is given as 50 miles, and the distance from A to B as 20 miles, it is evident that the distance from A to C is 50 miles minus 20 miles, or 30 miles. From illustrations similar to the above the pupils can formulate the rule that when one place is east and another is west of a given third place, to find the distance between the first two places we add their distances from the third place. When both are east or both west of the given third place, we subtract their distance from each other. In the illustration above, substitute the zero (Greenwich) meridian for city A, instead of city B substitute 20° west longitude, instead of city C substitute 30° east longitude, and we have the principle for determining when to add and when to subtract longitudes. To find the difference in longitude between two given places, if both places are east longitude or both are west, subtract the given longitudes; if one place is east longitude and the other is west, add the given longitudes. It is now customary to indicate west longitude as positive (+), and east longitude as negative (−). Problems similar to the following should be given to emphasize this principle:

City A is 95° west and city B is 30° east. What is their difference in longitude?

City A is 85° west and city B is 122° west. What is their difference in longitude?

City M is 120° east and city R is 18° east. What is their difference in longitude?

Illustrative Problems

If the course of study requires the solution of problems in longitude and time, in addition to those involving standard time, they may be solved as in the illustrations which follow. Frequent use should be made of a globe. The longitude of the city or the town in which the school is located can be approximately determined by careful measurements on a good map, and this longitude should be made the basis of a number of problems.

The longitude of Chicago, Illinois, is 87° 36′ 42″ W. The longitude of Washington, D. C., is 77° 2′ W. When it is 4 P. M. local time at Chicago, what is the local time at Washington?

The difference in longitude is (87° 36′42″) − (77° 2′) or 10° 34′ 42″.

Since 1° corresponds to 4 min.,
therefore 10° corresponds to 40 min.
Since 1′ corresponds to $\frac{1}{15}$ of a minute,
therefore 34′ corresponds to $2\frac{4}{15}$ min., or 2 min. 16 sec.
Since 1″ corresponds to $\frac{1}{15}$ of a second,
therefore 42″ corresponds to $2\frac{12}{15}$ sec., or $2\frac{4}{5}$ sec.

The total difference in time is 42 min. $18\frac{4}{5}$ sec.

Since Washington is east of Chicago, the time in Washington is later than in Chicago. The time in Washington is therefore 42 minutes $18\frac{4}{5}$ seconds after 4 P. M.

Pupils sometimes have difficulty in determining which of two places has the later time at a given moment. This difficulty may be eliminated by such questions as: Suppose it is noon where you live; is the sun also on the meridian of a city 30° west of where you live? How long before it will be on this meridian? The earth rotates from west to east; how long will it take a given point on the earth to rotate through 30°? By questions similar to this, the idea is developed that of any two places, the one farther west, measured on the shorter arc, has the slower time.

The longitude of New York City is 73° 58′ 26″ W., and that of Paris, France, is 2° 20′ 15″ E. When it is 3 :40 A. M. local time in New York City, what is the local time in Paris?

The difference in longitude is (73° 58′ 26″) + (2° 20′ 15″)

or 76° 18′ 41″. By use of the tables as in the preceding problem, we find that 76° 18′ 41″ corresponds to a difference in time of 5 hrs. 5 min. 14$\frac{11}{15}$ sec. Since Paris is east of New York, the local time in Paris is later than in New York. Therefore the time in Paris is 8 hrs. 45 min. 14$\frac{11}{15}$ sec. A. M.

Some text-books still use an incorrect form in solving a problem like the one above:

$$15)\overline{76 \qquad 18′ \qquad 41″}$$
$$\quad 5 \text{ hr.} \quad 5 \text{ min.} \quad 14\tfrac{11}{15} \text{ sec.}$$

Such a form is inaccurate and should not be tolerated. If a teacher advocates such a form because it is brief, he should recall the fact that brevity is not the thing to be chiefly sought in such work, and, further, that those whose business necessitates the solving of problems involving longitude and time, make their computations by the aid of longitude tables and not by the inaccurate method described above.

The longitude of Albany, New York, is 73° 44′ 48″. When it is 2 P. M. at Albany it is 20 min. 4 sec. past 1 P. M. at another city. Find the longitude of the second city.

The difference in time between the two cities is 39 min. 56 sec. A difference of 39 min. 56 sec. corresponds to a difference of 9° 59′ in longitude.

The second city must be west of Albany, since it has earlier local time. The longitude of the second city is (73° 44′ 48″) + (9° 59′) or 83° 43′ 48″. (This is the longitude of Ann Arbor, Michigan.)

Standard Time

After the relation between longitude and local time has been established, the subject of standard time should be considered. All places in the same longitude have exactly the same local time at any given moment. All places in different longitudes have different local time at any given moment. At the equator 1 minute of time corresponds to approximately 17 miles in longitude, and in latitude 40° to approximately 13 miles. It is evident that if every station along a railway extending east and west should keep its own local time there would be great confusion and the chance for accidents would be greatly increased. To avoid such confusion the railways of the United States in 1883 proposed a system of time that has since been adopted by most of the civilized nations of the world. The system is called "Standard Time" because the time at any given place is considered to be the same as the time at a certain standard meridian. In the United States these are four standard meridians. These meridians are the 75th, 90th, 105th, and 120th. The country (approximately $7\frac{1}{2}$° wide) on each side of these meridians is considered as a time belt, and every place in a given belt uses the time of its meridian. The belt which uses the time of the 75th meridian is called the eastern belt; the 90th meridian, the central belt; the 105th, the mountain belt; and the 120th meridian, the Pacific belt. It should be noted that the standard meridians are 15° apart. Since 15° correspond to 1 hour, it follows that the standard time in two consecutive time belts always differs by exactly one hour, the belt farther east always having the later time. The time belts are not exactly $7\frac{1}{2}$° wide on each side of the standard meridian. The belts were originally proposed by the railways for their own convenience, and the division terminals of the railways are not always just half way between two standard meridians. The line of division between two adjacent belts usually passes through the various railway terminals of the vicinity.

Every place in a given time belt has exactly the same

standard time as any other place in the same belt, and its time is just one hour earlier than the time of a place in the belt next to the east, and one hour later than that of any place in the belt next to the west.

By consulting the proper official railway guides, a teacher may find out the exact points where changes of time are made. The following are some of the principal division points between the time belts: Buffalo, New York; Pittsburgh, Pennsylvania; Wheeling and Huntington, West Virginia; Atlanta, Georgia; and Charleston, South Carolina, lie on the boundary between the Eastern and Central belts; while Mandan, South Dakota; North Platte, Nebraska; Dodge City, Kansas; and El Paso, Texas, lie on the boundary between the Central and Mountain belts.

The 75th meridian, which is the standard for the Eastern time belt, passes approximately through Philadelphia, Pennsylvania. The 90th meridian passes approximately through St. Louis; the 105th meridian passes through Denver; and the 120th meridian passes about 100 miles east of San Francisco.

Since the standard meridian of one of the time belts passes through Philadelphia, Pennsylvania, it follows that local and standard time are the same in that city. Since New York City is east of its standard meridian, the 75th, it follows that its local time is faster than it standard time; while in Harrisburg, Pennsylvania, which is west of its standard meridian, the 75th, the local time is slower than the standard time. Except where a time belt is very irregular, the extreme difference between local and standard time is about half an hour. If the exact difference between the local and the standard time in any locality is known, the longitude can be easily determined. Suppose the local time in your city is 12 min. 30 sec. faster than the standard time. Since 12 min. 30 sec. correspond to a differ-

ence of 3° 7′ 30″ of longitude, it follows that the city must
be 3° 7′ 30″ east of its standard meridian. (It must be
east of the standard meridian, since the local time is faster
than standard time.) If the city is in the Central time
belt its longitude is 90° − (3° 7′ 30″) or 86° 52′ 30″. If it
is in the Mountain time belt its longitude is 105° − (3°
7′ 30″) or 101° 52′ 30″. The pupils should have access
to a map showing the various time belts in the United
States.

The time belt whose principal meridian is 60° W. is
called the Atlantic belt. It embraces parts of Eastern
Canada. Great Britain, Holland, and Belgium use as their
standard meridian 0°. Most of the mid-European countries use the meridian 15° east as their standard. Bul-
garia, Roumania, and parts of Turkey use 30° east. South-
ern Australia and Japan use 135° east.

Problems in Standard Time

All problems involving standard time can be solved
orally. After the general boundaries of the time belts
in the United States are known, questions like the following
should be answered without difficulty:

When it is 10 A. M. Standard time in Omaha, Nebraska,
what is the time in Denver, New York City, Detroit, St.
Louis, San Francisco, and Ogden, Utah?

When it is 2 :35 P. M. Standard time in Indianapolis, what
is the time in New Haven; Portland, Oregon; Columbus,
Ohio; Austin, Texas; Salt Lake City; and Los Angeles?

Since longitudes in the United States are reckoned from
Greenwich, England, it is evident that the difference in
time between a place in England (0° meridian) and a place
in the United States is easily determined. Eastern time is
just five hours slower than Greenwich time. Central time

is six hours slower than Greenwich time, etc. When it is
10 A. M. at St. Louis, it is therefore six hours later, or 4 P.M.,
in London. When it is 11 P. M. Wednesday in St. Paul,
Minnesota, it is six hours later in London, England. The
time in London is therefore 5 A. M. Thursday.

The above consideration indicates how it is possible for
an event which happens at 2 P. M. in London to be given
in detail in newspapers in the United States which are
published at 10 A. M. of the same day.

Most pupils will be interested to determine the time in
Berlin, Germany; St. Petersburg, Russia; and Tokio,
Japan, when the school day begins in the United States.
Ask the pupils what time it is in Philadelphia, London,
New Orleans, San Francisco, Berlin, and Tokio at the
hour they are reciting.

The Date Line

The question of the international date line belongs primarily to geography, but it may be introduced here. The
time at that point on the earth's surface which is exactly
180° east of your school is 12 hours later than at your school,
since 180° corresponds to 12 hours, and since the time
must be later as the place is east. The time at that point
on the earth's surface which is exactly 180° west of your
school is 12 hours earlier that at your school, since 180°
corresponds to 12 hours, and since the place is west it must
have earlier time. But the place 180° east of the school
must be the same as the place 180° west of the school. It is
evident that at this place it cannot at the same instant be
both twelve hours earlier and twelve hours later than the
time at your school. This must be true, however, unless
some fact has been omitted in the calculation. In order to
rectify this apparent impossibility, another factor must be
taken into account; this factor is the date line.

Suppose that a man should start from your schoolhouse at noon on Wednesday and should travel due west with the same rapidity that the earth turns on its axis, or with the same rapidity that the sun appears to travel west. The sun would then be continually on the meridian of the traveler, hence for him the time would continually be Wednesday noon. He would make the trip around the earth in twenty-four hours, and when he reached the starting point the time to him would still be Wednesday noon, since the sun was continually on his meridian. To those who did not make the journey the time would be Thursday noon, since twenty-four hours elapsed and a night intervened. In order to make his reckoning of time correct the traveler would have to add one day. If we suppose the traveler to start at noon and travel east with the same rapidity as the earth rotates, or with the same rapidity that the sun appears to travel west, he would arrive at his starting point Friday noon, according to his reckoning, but Thursday noon according to the reckoning of those who did not make the journey. In order to rectify his dates the traveler must drop a day.

In the first illustration the traveler crossed the date line from the east towards the west; he had to add a day. In the second case the date line was crossed from the west towards the east; the traveler had to subtract a day. The same adjustment of dates would need to be made, no matter what the rates of travel might have been. The supposed rates make the illustration easy to follow.

The teacher should show the pupils that the dates of the traveler would have been correct had he added (or subtracted, as the case may be) a day at any point on his journey. His date upon arrival at the starting point would have been correct if he had added a day just after leaving the school or just before reaching it, or at any intermediate

point. This consideration leads to a discussion as to why
the international date line has its present location. The
teacher will find this point explained in any good mathematical geography. Space does not permit a fuller discussion of the question here than to state that most of the
islands whose earliest European inhabitants came by the
way of the Cape of Good Hope have the Asiatic date, while
those that were approached by way of Cape Horn have the
American date. This accounts for the irregularity of the
date line.

CHAPTER XXV

LITERAL ARITHMETIC AND ALGEBRA

There is a marked tendency in this country towards the introduction of some form of literal arithmetic or of algebra in the last two years of the grammar grades. In some schools the work in arithmetic is finished by the end of the seventh grade or by the end of the first half of the eighth, and the remaining time is devoted largely to algebra. In other schools some work in algebra is introduced into the seventh or eighth grades without dropping the arithmetic and with but little effort to relate the two subjects closely. In a few schools an effort is made to introduce literal arithmetic or algebra as opportunity permits while arithmetic is being studied, and an effort is made to establish a vital connection between the subjects. This last manner of introducing algebra is being rapidly extended and adopted. In about 35 per cent of the schools of the larger cities algebra is now taught in some form in the grades. Several arguments have been advanced to justify this practice. It is stated that a proper introduction of the subject interests the pupils to such an extent that they desire to enter the high school in order to extend their knowledge of it. Some advocate its introduction in order to make the transition from grammar school to high school subjects more gradual. Others introduce the simple equation, the graph, and algebraic formulas into the work in arithmetic in order to enrich the course and to simplify the solution of some of the more difficult problems.

The Present Tendency

The present tendency is towards the introduction of some work in algebraic formulas, simple equations, and graphs in connection with the regular work in arithmetic. There is an attempt to break down the traditional demarcations between arithmetic and algebra and to correlate the subjects more closely. Many pupils of the upper grammar grades do not enter the high school, and these pupils should have the opportunity to learn the meaning and the use of algebraic formulas and graphs, now so frequently used in trade journals and mechanics' handbooks, and to appreciate the value of the simple equation as a tool in the solution of problems.

One of the chief objections that has been urged against algebra, as it is usually taught in the grammar schools, is that it is an attempt to complete a definite portion of the work of the first year of the high school. There is often too much emphasis upon definitions, theory, and complicated processes. A pupil who does not enter the high school derives but little benefit from a formal study of the four fundamental operations and factoring in algebra.

In the latter part of the sixth and in the seventh and eighth grades the teacher should not hesitate to introduce letters to represent numbers. It is no more unnatural, after a short time, for the pupil to use "C" to represent the cost, "I" to represent interest, "V" to represent volume, and "N" to represent a number than for him to use N.Y. to represent New York or A.M. to indicate time before noon. Pupils soon appreciate the fact that calculations involving letters are often simpler than those involving figures. As Young says, "The use of letters to represent numbers simplifies the treatment of some types of problems which otherwise tend to confuse by

mere verbiage.''[1] If only those processes that are actually
needed in the solution of the problems of arithmetic are
emphasized, but little objection will be offered in most
communities to the introduction of such work in the grades.
The transformations that are necessary in the solution of
simple equations and in the evaluation of simple algebraic
formulas can be readily understood by pupils of the seventh
and eighth-grades.

Strong protests are always made when any innovations
in subject-matter or symbols are suggested. Improvements
are usually made in spite of opposition and protest. Smith
has pointed out the fact that when it was proposed to write
''4×5'' instead of ''4 times 5'' strong protests were made
on the ground that ''4×5'' was more abstract than ''4
times 5'' and that it would be best to let well enough
alone.[2] The pupil should recognize the symbols of algebra
as a shorthand method of indicating magnitudes. Arithmetic does not become algebra by the mere use of letters
instead of numbers. The solution of many of the problems
of common and decimal fractions, of ratio and proportion, percentage and mensuration, is greatly simplified and
abridged by the use of letters for the magnitudes which
they represent.

The pupil who understands the following forms is prepared to understand the fundamental principles of simple
equations:

$$4 + ? = 6 \qquad\qquad 3 \times ? = 12$$
$$8 - ? = 5 \qquad\qquad 12 \div ? = 6$$
$$? + 3 = 7 \qquad\qquad ? \times 4 = 12$$
$$? - 5 = 4 \qquad\qquad ? \div 4 = 8$$

If instead of the interrogation point the symbol ''n'' is

---

[1] Young, ''The Teaching of Mathematics,'' p. 243.
[2] Smith, ''The Teaching of Arithmetic,'' pp. 72-73.

used to represent the required number, its value may be readily found. If teachers will refrain from the introduction of unnecessary definitions and emphasis upon technical and non-essential points, they will find that pupils like to use the literal notation. In this, as in other subjects, teachers often befog and obscure what is otherwise clear to the pupil by attempting to give minute explanations.

The pupil should appreciate the fact that an algebraic formula is always an abridged and concise method of stating a fact, and he should be able to translate a formula into words and to translate a simple statement of condition into a formula. If "l" represents the length of a field in rods and "b" represents its breadth in rods, what does $lb = 1600$ state? If "v" represents the volume of a sphere and "r" represents its radius, what does $v = \frac{4}{3}\pi r^3$ state? Such expressions as those above should be as intelligible to the pupil when stated by the use of letters as when stated in words. The literal statement of a problem is more compact than the statement in words, and the relations between the various factors involved may usually be more clearly seen.

The value of the equation as a tool in the solution of problems may be impressed upon the pupil by requiring him to solve some of the problems of the text without the use of letters, and then comparing with these the solutions of the same problems in which literal notation and the simple equation have been employed. Such comparisons are often a revelation to a pupil.

CHAPTER XXVI

PRESENT TENDENCIES IN ARITHMETIC

A few decades ago the teachers of arithmetic made little effort to appeal to the pupil's interests by utilizing his experiences. Memory was an important factor in most school work, and drill was an essential part of class instruction. To-day we recognize that the value of a study to a pupil depends largely upon the amount of mental energy that the pupil puts into it, and that this in turn is largely dependent upon the interest with which the pupil pursues the study. Interest in a subject lies very near the basis for success in the subject. That a child learns through his experiences is one of the central facts of modern pedagogy; and as this fact meets with more general acceptance, increasing emphasis will be placed upon the child's own activities.

The pupil of the lower elementary school is more or less a creature of the present. His dominant interests are in things that appeal to him because of immediate utility or pleasure. The strongest motives for good work in the lower grades are based upon the pupil's dominant interest at the time. As the pupil matures and his educational horizon broadens, his interest may be aroused by the use of motives more or less remote in time. It is the duty of the teacher to arouse interest in the subject and to utilize this interest to secure that careful and consistent study which is a prerequisite of the best educational results. Persistent application is the price that must be paid for successful mastery of any subject.

The School as a Social Institution

The school is gaining recognition as a social institution and we are beginning to realize that social efficiency means other than mere business efficiency. The pupil has the right to be informed in regard to the broader aspects of modern social, industrial, and commercial activities, and it is the duty of the school to see that he acquires this information. In so far as this information involves the larger quantitative aspects of those activities it may properly be included in a course in arithmetic. "Mind furnishing and mind training go hand in hand."

Problem Material

One of the marked features of the arithmetic of to-day is the attempt to adapt the problem material to the needs and interests of the pupil instead of adapting the pupil to the problem, as was frequently attempted by the older texts. It must be admitted that the result is often a loss to the pupil. Numerous problems relating to the common phases of community life are being introduced. The aim is to secure the maximum amount of self-activity on the part of the pupil by confronting him with problems which appeal to him as concrete and vital. There is a growing recognition of the fact that there should be a legitimate motive or purpose underlying a problem, and that as many problems as possible should be more or less related to matters that are within the experience of the pupil. A problem that appeals to an adult as real and vital may not make the same appeal to a child. A problem that appeals strongly to a pupil of the fifth or sixth grade may be of little value to the pupil of the second or third grade, even with smaller numbers; and some of the number games of the lower grade would be out of place in the higher grades. The appeal should be to the interests and activities of the pupil in and out of school, and the interests of the adult should be regarded as of subordinate importance. A problem is not of maximum value merely because it is about a factory, a store, or a bank. It should be of a type that is actually met by those who do the world's work, and the data involved should be within the intelligence and the experience of the pupil. A problem may be concrete and full of significance to one pupil and not at all so to another pupil. Myers has pointed out the fact that "children's problems are not merely men's or women's problems cracked up into smaller bits. They must differ qualitatively as well as quantitatively."[1] Most pupils are anxious to solve problems that actually come within their own experience. The fact that a pupil is interested to know how to solve a problem does not necessarily imply that he will be able to solve it, but much has been gained when problems have been so chosen that pupils are willing and eager to learn how to solve them. Interest begets effort, and effort properly directed produces results. Any problem that appeals to the pupil may legitimately be used in arithmetic, provided it does not give a false idea of social, industrial, and commercial activities of the day. Text-books suggest numerous types of problems, and the teacher should supplement them by problems of a local character. Pupils should be encouraged to bring in problems that appeal to them as interesting.

Many texts of former years went to the extreme of devoting too little attention to the applications of arithmetic, and the reaction against this tendency is likely to carry us too far in the other direction. Many problems dealing with the various relations and properties of numbers are

[1] "Arithmetic in Public Education," Myers, pp. 8-9.

more interesting to the pupil than certain problems of the shop. The essential thing is that the problem shall really appeal to the pupil as interesting. To use only those problems that involve buying and selling, measuring and estimating, and the various business procedures, is to eliminate from arithmetic many problems that are intensely interesting to the pupils. It is as unwise to require pupils to solve numerous problems involving applications of which they have no conception as to eliminate all of the interesting problems of former years merely because they find no immediate application in present-day activities. Those who have adopted the extreme utilitarian view and insist that nothing be taught except that which "functions in the immediate present" are not numerous, but their influence is quite widely felt. A pupil may become intensely interested in solving a problem which appeals to his fancy, even though it is in no way related to the so-called practical. In the broader sense of the term, any problem in which the pupil is really interested, and which does not give him false ideas, is practical for him, because it helps to develop his ability to understand number relations.

Some enthusiasts have introduced into the seventh and eighth grades numerous problems that are social in content but involve nothing but the mathematics of the primary school. The introduction of problems of this type may be quite valuable in classes in elementary sociology and economics, but such problems do not contribute to mathematical insight or skill.

We must try to make arithmetic interesting to the pupils, and a judicious selection of problems is one means of accomplishing this end. However, practical utility is not the only criterion by which the worth of a problem should be judged. One of the keenest pleasures a pupil experiences in the study of arithmetic comes from the ability to overcome the difficulties involved in the solution of problems. The joy of conquest is a large factor in the pupil's intellectual life. Too often pupils experience little of the pleasure that comes from the successful mastery of a difficult task: The joy of intellectual effort and of mental acquisition are too little known by the modern pupil. Patience and persistency ·of effort are not sufficiently emphasized.

It is difficult to secure an adequate supply of practical problems involving the mathematical principles which must be mastered, but great advance has been made in recent years in this respect, and no doubt further search will reveal other types of problems of the desired kind. At present, however, a pupil cannot acquire the desired facility and accuracy in mathematical operations and develop sufficiently his power to discriminate the essential relations of the various elements involved if the problems are restricted to the utilitarian types that are immediately within his experience. Furthermore, it is not true that the pupil's only interest is in this type of problem. He may become interested in the solution of a problem that has no immediate relation to utilitarian ends; one that cannot be directly or indirectly correlated with the practical. If, however, the problem makes a strong appeal to him; if it engages his attention and challenges his powers, it is serving the purpose. A problem that appeals to the pupil "just for the joy of the doing" must not be left out, even though it has no other value. Such problems serve a useful purpose and are too few.

Not every task in life or in school is a pleasant one at the time it is being performed. We may wish that it were so and may strive to make it so, but the fact remains that some things must be done both in and out of school that are not in themselves interesting. Such work may have a very direct bearing on other work that will be interesting because of the mastery of something in itself uninteresting. The thoughtful teacher welcomes every suggestion that makes any school activity more interesting and is continually seeking for ways and means to accomplish this end, but no way has yet been discovered to make all tasks interesting to all the pupils all the time. Our pupils need to have developed within them the habit of sticking to a task until it is successfully completed, even though the doing of the work may not appeal to them as interesting. No good teacher will go to the extreme of giving pupils an uninteresting task to perform merely because it is uninteresting, but it is almost as great an error to eliminate everything that does not immediately engage the pupil's interest. The teacher who is interested in his work usually finds that his enthusiasm in the subject is contagious among his pupils. If the teacher increases his margin of knowledge and becomes a master of the field that he is teaching, he usually finds that interest begets interest, and that the more he puts into his subject, the more response he gets from his pupils and the more interest is aroused. Much grind is due to the fact that the teacher has neither the ability nor the energy to enliven the subject or to awaken interest and enthusiasm by being interested and enthusiastic himself.

Omission and Introduction

The tendency to-day is to omit from arithmetic any topic that time or changing social conditions has rendered obsolete or purely technical for a small group, and to revitalize the topics that remain so that they will represent actual conditions of the present. There is increasing emphasis upon the fact that school work in arithmetic should prepare the pupil to deal with "out of school problems.'' Arithmetic should train the pupil to see the world from a quantitative point of view. Emphasis is placed upon the types of problems that frequently occur in the larger "out of school" activities, and no attempt is made to present every kind and form of problem that may be practical in minor trades in which the pupils have no immediate concern. The school must give the pupil a quantitative knowledge of those facts which he must some day know in order to assume his place in social and industrial life. Problems of food and clothing supply, of transportation, of buying and selling, of building, of mining, of city and county administration are emphasized.

Problems of the farm are receiving greater emphasis in rural communities, and problems involving the numerous activities of city life are emphasized in urban communities. Well-executed and accurate pictures which really aid in the understanding of number relations are being used increasingly in text-books.

Unduly long and complicated problems are being omitted and greater emphasis is being placed upon the essentials of arithmetic. "There is a growing belief that the aim of the work in arithmetic should be limited to accuracy and a reasonable facility in the fundamental operations—addition, subtraction, multiplication, and division of whole numbers and simple fractions; and to *simple practical problems* involving the operations together with some instruction in percentage and its simplest applications to interest, trade discount, taxes, and insurance."[1]

"The obsolete and the relatively infrequent, the overcomplex and the wasteful processes of the old arithmetic tend to disappear."[2]

[1] Paul H. Hanus, Harvard College, in a Report on the Program of Studies, 1911.
[2] Suzzalo, "The Teaching of Primary Arithmetic."

The time saved by the omission of obsolete topics and the abridged treatment of some others is being utilized in a number of ways. We are beginning to appreciate the extent to which arithmetic may be made to contribute to social insight. Pupils are shown how the world uses its mathematics by visits to centers of commercial and industrial activity and the setting of the problem and its solution are studied. Instead of devoting all of the time to the solution of problems about a factory or a bank, the modern class occasionally visits these institutions. A good course in arithmetic to-day includes a consideration of the following topics, all of which have been introduced recently: The saving and loaning of money; the investing of money; modern banking methods; keeping of simple accounts; a study of tax levies; public expenditures and insurance from the social point of view. Much of the work in these topics must be of the informational character, and belongs quite as much to a course in civics and economics as to arithmetic.

The various factors which make for the socialization of arithmetic will, if properly controlled and judiciously used, increase the pupil's efficiency in the subject. Many teachers will carry the idea to absurd extremes, and as a result their pupils will have but little mastery of the essential number facts. Any good tendency may become bad when carried too far.

Skill in Computation

The arithmetic of to-day is generally less abstract and formal than that of a few decades ago. Manual training, industrial and household arts, geography, history, agriculture, and elementary physics are being utilized as sources of problems. Arithmetic is less a series of abstract exercises and more a tool useful in other subjects. The fact that much of the work is being motived does not mean that the drill, so necessary to fix in mind the fundamental number facts, should be relegated to a subordinate position. The more arithmetic is used in practical ways, the more must the pupils realize the need of mastery of the fundamental number relations; and this in itself is one motive for drill. There is increasing emphasis to-day on systematic drill.[1] There is a growing belief in the necessity of concentrating the attention at times upon the operations alone. If our pupils are to learn to compute with facility and accuracy, they must have a great deal of practice in computing. Arithmetic is both a science and an art. Exercises are necessary to develop skill.

The Solution of Problems

When a pupil is required to solve a problem, he must first comprehend the situation presented and must then decide upon the operations to be used, then the actual computations must be made. A large number of examples must be given in order to develop skill in computation. By the time the pupil has reached the latter grades of the grammar school the fundamental operations ought to be so thoroughly mastered that his mind is free to devote itself to the solution of the problem rather than to the performance of the operation. The solution of a problem involves knowing what to do, and the doing of it should be a minor factor in the upper grades. Pupils need a great deal of drill in interpreting problems; this phase of the work has received too little emphasis in most schools. It is wise to require a pupil frequently to transform a "problem" into an "example"; that is, to require him to analyze the data given in the problem, in order to

[1] See chapter on Drill.

determine the proper processes to be used, writing down in one column the analysis in systematic form, and in another column doing the necessary computing, using all the short cuts possible. In the one column, emphasis is put upon accuracy of thought processes, in the other, on accuracy of computation. The modern course of study limits the problems largely to matter coming within the pupil's experience, and the modern school patron is demanding efficiency within this narrowed field. The solution of a large number of problems of medium difficulty is of more value to the pupil than the solution of a few problems of great difficulty. The teacher of to-day is placing emphasis upon oral arithmetic, and we are returning to a reasonable use of that important phase of·the work.

The Analysis of Problems

Not only is there a tendency to master the quantitative side of life by utilizing the pupil's experience and by correlating the work in arithmetic with the work in other subjects, but there is a decided tendency to foster a spirit of inquiry and to develop the power to interpret number relations. Instead of requiring a pupil to solve a problem according to some type form, we seek to encourage his originality and individuality by permitting more flexibility in analysis than was formerly allowed. We encourage him to choose the method that seems best to him, and then we ask him, at times, to justify his choice. The rigidity of full logical form in analysis is not considered so important as in prevous years. A pupil who always associates a particular process with specific words of relation used in a problem is often unable to solve the problem when the phraseology is even slightly changed. Marked variation in the difficulty of a problem may be caused by a difference of phraseology. By a judicious lack of uniformity in the phraseology we force the pupil to rely upon his own resources in solving them. The pupil is encouraged to see the relation between what is given and what is required, and to choose the appropriate process. Superficial and thoughtless associations are discouraged.

The difficulty in a given problem usually lies in the fact that the pupil is not able to break the problem up into its various steps and to apprehend the order of the steps. Unless he has more or less ability to analyze problems, his knowledge of arithmetic will be more or less mechanical. The pupil who continually depends upon rules or type problems has attained but little mastery of the subject. He should be encouraged to devise solutions of his own. The more insight he has into the relationships involved in a given problem, the greater freedom will he have from mechanical procedure.

Rationalizing the Processes

There is a marked tendency to-day to rationalize the processes in arithmetic.[1] It is not assumed that the pupil will be able to give a clear-cut and logical explanation of every process, but each process should be developed and explained in such a way that the entire procedure appeals to him as reasonable. For example, instead of being told to invert the divisor in division of fractions and multiply by the dividend, he is first shown why such a procedure will give the correct result.

Unity of Arithmetic

Another tendency in the teaching of arithmetic is the effort to develop the subject as a whole rather than as a

[1] See chapter on Rules and Analysis.

series of more or less unrelated topics. More emphasis is
being placed to-day on the mastery of principles and less
on the mastery of rules and definitions than was the case
in former years. The pupil who fails to grasp the few
simple underlying principles of the subject has seen but
little of the beauty and the simplicity of arithmetic. The
number of distinct mathematical principles involved is sur-
prisingly small, but unless a pupil grasps these principles
the whole subject lacks organic unity in his mind. It is
important to analyze and to classify topics; to be able to
distinguish points of similarity and of difference; to develop the power to see relations between topics that are
based on the same underlying principles. As McMurry
says, ''Merely to go through a text-book without picking
up the strings and tying them together is to fail in a most
essential thing.''[1] The pupil who does not see the unity
in arithmetic fancies that he is dealing with something
entirely new from the mathematical point of view every
time he takes up a new topic, whereas he may be dealing
with a new phase of a topic long familiar to him. Real
knowledge implies organization, and organization, in turn,
implies more or less unity. Spencer says, ''When a man's
knowledge is not in order, the more of it he has the greater
will be his confusion of thought. When the facts are not
organized into faculty, the greater the mass of them the
more will the mind struggle under its burden, hampered
instead of helped by its acquisition.''

There is a tendency to-day not only to unify the subject as a whole, but to introduce more or less unity into
the assignment of problems for a lesson or a group of
lessons. The custom has been to group problems more or
less into series involving the same processes but different
in all other respects. The present tendency is to ''attain a

[1] See McMurry, ''How to Study,'' p. 79.

more approximate unity within the subject-matter of the problems themselves.''

Formal Definitions

There is a marked tendency to-day to minimize the importance of formal definitions, especially in the lower grades. The meaning of each term used should be clearly understood, but it is not wise to require formal definitions in arithmetic from young children. A pupil in the lower grades, if required to give a formal definition of an arithmetical term, usually repeats the exact words of the textbook, and the real meaning is but little understood. Too often the pupil who states a definition most glibly is repeating words without expressing ideas. It is probable that no child gets the idea of number from a definition of it. In the later grammar grades some attention may properly be given to the definition of terms, but no attempt should be made to define a term until it has been illustrated. Pupils should be trained to formulate, the few definitions required, in a clear and concise manner, and the teacher should determine by repeated illustrations whether the definition has the proper content in the pupil's mind. The few definitions that are learned should be expressed in clear and forceful language. The attempt to-day in arithmetic is to teach the pupils ''to observe, to think, to do, rather than to repeat and to memorize.''[1]

Scientific vs. Empirical Basis

Another tendency in arithmetic is the attempt to establish the various methods of presenting the subject upon a scientific rather than an empirical basis. Serious attempts

[1] Howland.

are being made to determine by means of controlled experiments the relative value of the methods that are generally
advocated, and to substitute verified conclusions for individual opinion. In these days of educational unrest and
reform it is not strange that advocates of extreme methods
should appear. Not every suggestion for reform is in the
right direction, and it is as necessary to reject new errors
as it is to eliminate old ones. ''We must not lose our
bearings in the midst of the unscientific radicalism of the
day.'' We must examine with great care that which the
pedagogic alchemist would have us try. We must follow
the conclusion of our best thoughts and not be bound by
the prejudice of years of practice. The basis for any
successful system of pedagogy must be laid in the expe-
ricuce of preceding generations. It is necessary to master
the pedagogy of the mathematics of the past, its aims, its
methods, its devices, in order to estimate the advantages
of previous attainments and to adapt the best to the needs
of our generation. The teacher who knows the history of
his subject is able to make a conservative selection among
the many radical suggestions and schemes for improvement, each of which probably has a foundation of truth
but is given an exaggerated importance by its chief advocate. There must be a careful weighing and adjusting.
Through a knowledge of the history of arithmetic we gain
an ability to deal with the modern suggestions for improvement which we shall be called upon to judge. We also
avoid the ridiculous situation of advocating as new a
method or device proposed and discarded centuries ago.

Investigations in Arithmetic

In the chapter on Investigations in Arithmetic the conclusions of the noteworthy scientific experiments that have
been made are briefly enumerated. More of our teaching

is based upon the results of such investigation to-day than ever before in the history of education, and more studies are imperatively needed in every field of school work. There has been too much tacit acquiescence to general opinion in education. Not every teacher has the ability or the training that justifies him in attempting to plan and to conduct an experiment to determine the relative values of different methods of procedure. Numerous factors may enter into the investigation, and unless these are given proper weight the entire experiment may be of no value. Most teachers, however, can coöperate with an investigator of ability, and can thus aid in increasing the amount of knowledge based on controlled experiment.

Text-books

Another tendency in the arithmetic of to-day is seen in the organization of the material of the text-book and the attitude of the teacher toward the texts. Some years ago, when the faculty psychology was in vogue, there was extreme regard for logical arrangement. The disciplinary side of arithmetic was emphasized much more than the practical side. Gradually the demands for the elimination of the obsolete and the impractical became more insistent, and we are witnessing the results of this movement. The old arithmetic sought to give a complete treatment of each topic before proceeding to the next. The rise of the kindergarten movement destroyed this ideal, for it was soon recognized as impossible to give a thorough treatment of any topic to a child of five, if his interests and capacities were considered. The Grube method was a prominent influence against the extreme topical plan, but it was just as logically impossible. The eagerness with which the method was adopted indicated the desire to discard the extreme topical plan. The wearisome monotony of the Grube method and the impossible thoroughness which was sought tended to destroy the child's interest in number.

The Spiral Plan

A direct outgrowth of the reaction against the extreme topical plan and the Grube method was the Spiral Plan. It sought to continue the Grube method in part, but eliminated the idea of extreme thoroughness. The plan met with wide acceptance, and numerous texts based on this idea appeared. As more topics were introduced, the coils of the spiral shortened and the pupils were nauseated by the frequent recurrence of the same topics. The reaction against the extreme spiral form set in, and the trend is now towards much longer spirals. In many sections of the country there is a very strong sentiment against any form of spirals. The best texts of to-day use neither the extreme spiral nor the extreme topical arrangement of subjects, but attempt to use the best features of each plan. The ideal arrangement is probably somewhere between the two extremes.

Influence of Text-books on Teaching

In 1895 a book entitled "The Psychology of Number" appeared. The authors, McClellan and Dewey, advocated the idea of teaching arithmetic by placing great emphasis on measurement as the basis of number ideas. The book exerted a marked influence on teachers, and many courses were organized about measurement as the basis of number concepts.

Text-books influence the teaching of arithmetic to a great degree. Thousands of teachers are largely dependent upon text-books for their method, organization, material, and sequence of topics. Originality and initiative are as rare among teachers as among individuals in other vocations and professions. Most teachers are followers, not leaders. There is to-day a growing independence among teachers in regard to text-books. Many of the older texts were followed slavishly, and a few of them even indicated the questions that the teacher was to ask and specified the pages on which the answer was to be found. There is a growing recognition of the fact that no book can adequately meet the needs of any class at every instant. It is the duty of the teacher to introduce numerous exercises not in the text, to give many of the problems a local setting, and to omit such topics and problems as are not adapted to the needs and capacities of the pupils. It was formerly regarded as a confession of weakness for a teacher to omit any topic or any problem found in the text. We are beginning to recognize that a judicious elimination and supplementing of the material of the text-books is an indication of strength. The text-book should not be the teacher's creed. Many teachers are prone to view with suspicion any method of procedure that a pupil ventures to introduce unless it corresponds with the method of the text. Such an attitude towards a text-book tends to repress rather than to encourage that spirit of inquiry and of originality which arithmetic should seek to cultivate. The use of a text-book is, in most cases, economical both of time and of energy. The text-book should be the basis for the work. A good teacher will get good results in spite of a poor text-book, but a well-written text increases the efficiency of the teacher. A poor teacher will obtain poor results from any text-book.

Most authors of text-books are followers rather than leaders in educational thought. If an author is too radical or too conservative in his material or method, the sale of his book is greatly curtailed.

Preparation of a Course of Study

The recent tendency of many teachers to break away entirely from written texts and to prepare courses of study in arithmetic which they think are better adapted to their localities, are not to be commended. A good course of study involves a large number of factors, and the old maxim that everybody knows better than anybody is particularly applicable here. We should not lightly discard the accumulated wisdom of the past and strive to build up a course of study solely from our own theories. Not everything that was done in the schools of other years was bad, and not everything was good, but to discard both good and bad is folly. To prepare a good course of study in arithmetic is a very difficult task. It demands not only a knowledge of local conditions where the course is to be used, but it demands a knowledge of the history of the subject. It demands a breadth of view and a sanity of judgment that not many possess. It demands a knowledge of the best that is being done in other countries. It demands clear ideas as to the purpose for which arithmetic should be taught in the schools. Local adaptations are comparatively easy to make, but a good course of study is rare indeed.

CHAPTER XXVII

COURSES OF STUDY

One of the most satisfactory studies of a local situation as a factor in determining the character of a course of study in arithmetic that has come to our notice was made* by Mr. G. M. Wilson, formerly Superintendent of Schools at Connersville, Indiana. The problem was attacked (1) through grade meetings, (2) by comparing forty-seven representative courses of study, and (3) by securing the testimony and assistance of local business men.

The course of study resulting from this investigation is organized upon the basis of social utility. Obsolete and complicated materials are eliminated. The arithmetic of new forms of business practice are introduced. The simplest forms of solution are encouraged.

The grade occurrence of arithmetic topics as shown by these forty-seven courses of study was:

| | GRADES | | | | | | | |
Subject	I	II	III	IV	V	VI	VII	VIII
Numeration	37	37	33	18	7	4	4	1
Notation	34	39	33	18	7	4	4	1
Relation of Numbers	10	11	6	5	5	4	4	3
Addition	23	39	33	28	13	9	6	3
Subtraction	24	39	34	26	13	9	6	3
Multiplication	10	28	30	25	15	15	10	10
Division	2	16	21	34	27	25	19	17
Fractions	10	28	22	31	34	28	24	24
Denominate Numbers	11	20	23	34	30	31	29	22
Involution and Evolution	..	..	..	..	1	1	7	16

GRADES—*Continued*

Subject	I	II	III	IV	V	VI	VII	VIII
Decimal Fractions.........	..	..	..	5	23	12	8	1
Mensuration	6	8	10	11	12	14	14	12
Multiplication Tables.....	2	8	20	18	7	5	7	6
Commission and Brokerage	..	..	..	..	..	10	11	6
Insurance	..	..	..	..	..	10	9	6
Percentage	..	..	..	..	7	16	13	8
Ratio and Proportion......	..	..	..	1	3	8	6	9
Partnership	..	..	..	..	..	2	7	4
Partial Payments.........	..	..	..	..	..	..	3	5
G. C. D. and L. C. M.....	..	..	..	..	4	6	..	..
Longitude and Time.......	..	..	..	..	..	4	7	1
Profit and Loss..........	..	..	..	..	..	7	17	2
Taxes	..	..	..	..	..	2	14	3
Duties	..	..	..	..	..	1	13	1
Banking	..	..	..	..	..	..	9	6
Exchange	..	..	..	..	..	..	4	4
Simple Interest..........	..	..	..	1	2	12	23	4
Stocks and Bonds........	..	..	..	..	..	..	6	8
Business Forms..........	..	..	..	..	1	4	15	6
Simple Accounts.........	..	..	..	3	6	5	3	3

SOME VALUABLE BOOKS FOR SUPERVISORS AND TEACHERS OF ARITHMETIC

"History of Mathematics," Ball; The Macmillan Company, Chicago.

"History of Elementary Mathematics," Cajorie; The Macmillan Company, Chicago.

"A Brief History of Mathematics," Fink; The Open Court Publishing Company, Chicago.

"The Educational Significance of Sixteenth Century Arithmetic," Jackson; Teachers' College Bureau of Publications, New York.

"Rara Arithmetica," Smith; Ginn & Company, New York.

"The Hindu-Arabic Numerals," Smith and Karpinski; Ginn & Company, New York.

"The Psychology of Number," McClellan & Dewey; D. Appleton & Company, Chicago.

"A Fundamental Study in the Pedagogy of Arithmetic," Howell; The Macmillan Company, Chicago.

"The Number Concept," Conant; The Macmillan Company, Chicago.

"The Teaching of Elementary Mathematics," Smith; Ginn & Company, New York.

"The Teaching of Mathematics," Young; Longmans, Green & Company, New York.

"The Teaching of Arithmetic," Smith; Ginn & Company, New York.

"Methods in Arithmetic," Walsh; D. C. Heath & Company, New York.

"The Teaching of Arithmetic," Stamper; American Book Company, Cincinnati.

"Special Method in Arithmetic," McMurry; The Macmillan Company, Chicago.

"Primary Arithmetic," Suzzallo; Houghton, Mifflin Company, Boston.

"Arithmetical Abilities and Some Factors which Determine Them," Stone; Teachers' College Bureau of Publication, New York.

"Standard Tests in Arithmetic," Courtis. [A series of tests prepared by and for sale by S. A. Courtis, Detroit, Mich.]

"Practice in the Case of School Children," Kirby; Teachers' College Bureau of Publication, New York.

"Arithmetic Supervision," Jessup & Coffman; The Macmillan Company, Chicago.

"Number Rhymes and Number Games," Smith; Teachers' College Bureau of Publication, New York.

Report of the American Commissioners on "The Teaching of Mathematics in the Elementary Schools," Bureau of Education, Washington, D. C.

"How to Study and What to Study," Sandwick; D. C. Heath & Company, Boston.

RECREATIONS

"Mathematical Recreations," Ball; The Macmillan Company, Chicago.

"A Scrap Book of Mathematics," White; The Open Court Publishing Company, Chicago.

"Mathematical Wrinkles," Jones; S. I. Jones, Gunter, Texas.

"The Canterbury Puzzles," Dudeney; E. P. Dutton & Company, New York.

INDEX

PRICE LIST

Essential Studies in English, *Robbins and Row.*
    Book I, Language, 294 pp..........................$0.45
    Book II, Grammar and Composition, 356 pp........ .60
Practical English, *A. C. Scott,* 208 pp................. .60
Manual of English Form and Diction, *Fansler*........... .10
Exercises in English Form and Diction, *Fansler*........ .60
Types of Prose Narrative, *Fansler*.................... 1.50
A Practical Spelling Book............................ .20
The National Speller, *C. R. Frazier*.................... .20
Phonology and Orthoepy, *Salisbury*.................... .50
Elementary Agriculture, *Hatch and Haselwood*......... .50
A Unit of Agriculture, *Eliff*........................... .50
One Hundred Lessons in Agriculture, *Nolan*........... .65
The Educational Meaning of Manual Arts, *Row*......... 1.25
Methods of Teaching, *Charters.* Revised Edition....... 1.25
Principles of Teaching, *N. A. Harvey,* 450 pp........... 1.25
The Theory of Teaching, *Salisbury*.................... 1.25
Reading in Public Schools, *Briggs and Coffman*........ 1.25
The Psychology of Conduct, *Schroeder*................. 1.25
The Personality of the Teacher, *McKenny*.............. 1.00
Country Life and the Country School, *Carney*.......... 1.25
School Management, *Salisbury*........................ 1.00
Index to Short Stories, *Salisbury and Beckwith*......... .50
Balonglong, the Igorot Boy, *Jenks*.................... .45
Reading-Literature Primer, *Free and Treadwell*........ .32
Reading-Literature First Reader....................... .36
Reading-Literature Second Reader..................... .40
Reading-Literature Third Reader...................... .45
Reading-Literature Fourth Reader..................... .50
Reading-Literature Fifth Reader...................... .55
Reading-Literature Sixth Reader. *Ready in May, 1914.*
Reading-Literature Seventh Reader. *Ready in May, 1914.*
Reading-Literature Eighth Reader, *Shryock*............ .60
Ivanhoe, adapted by *Stella Humphrey Nida*............ .50
East O' the Sun and West O' the Moon, *Thomsen*...... .50

ROW, PETERSON & COMPANY
623 So. Wabash Avenue   **CHICAGO**

Lightning Source UK Ltd.
Milton Keynes UK
UKHW042357031118
331733UK00008B/1749/P